Contributions of the Committee on Desert and Arid Zones Research of the Southwestern and Rocky Mountain Division of the American Association for the Advancement of Science

Previous Symposia of the Series

1. *Climate and Man in the Southwest.* University of Arizona, Tucson, Arizona. Terah L. Smiley, editor. 1957
2. *Bioecology of the Arid and Semiarid Lands of the Southwest.* New Mexico Highlands University, Las Vegas, New Mexico. Lora M. Shields and J. Linton Gardner, editors. 1958
3. *Agricultural Problems in Arid and Semiarid Environments.* University of Wyoming, Laramie, Wyoming. Alan A. Bettle, editor. 1959
4. *Water Yield in Relation to Environment in the Southwestern United States.* Sul Ross College, Alpine, Texas. Barton H. Warnock and J. Linton Gardner, editors. 1960
5. *Ecology of Groundwater in the Southwestern United States.* Arizona State University, Tempe, Arizona. Joel E. Fletcher, editor. 1961
6. *Water Improvement.* American Association for the Advancement of Science, Denver, Colorado. J. A. Schufle and Joel E. Fletcher, editors. 1961
7. *Indian and Spanish-American Adjustments to Arid and Semiarid Environments.* Texas Technological College, Lubbock, Texas. Clark S. Knowlton, editor. 1964
8. *Native Plants and Animals as Resources in Arid Land of the Southwestern United States.* Arizona State College, Flagstaff, Arizona. Gordon L. Bender, editor. 1965
9. *Social Research in North American Moisture-Deficient Regions.* New Mexico State University, Las Cruces, New Mexico. John W. Bennett, editor. 1966
10. *Water Supplies for Arid Regions.* University of Arizona, Tucson, Arizona. J. Linton Gardner and Lloyd E. Myers, editors. 1967
11. *International Water Law Along the Mexican-American Border.* University of Texas at El Paso, Texas. Clark S. Knowlton, editor. 1968
12. *Future Environments of Arid Regions of the Southwest.* Colorado College, Colorado Springs, Colorado. Gordon L. Bender, editor. 1969
13. *Saline Water.* American Association for the Advancement of Science, New Mexico Highlands University, Las Vegas, New Mexico. Richard B. Mattox, editor. 1970
14. *Health Related Problems in Arid Lands.* Arizona State University, Tempe, Arizona. M. L. Riedesel, editor. 1971
15. *The High Plains: Problems in a Semiarid Environment.* Colorado State University, Fort Collins, Colorado. Donald D. MacPhail, editor. 1972
16. *Responses to the Dilemma: Environmental Quality vs. Economic Development.* Texas Tech University, Lubbock, Texas. William A. Dick-Peddie, editor. 1973
17. *The Reclamation of Disturbed Arid Lands.* American Association for the Advancement of Science, Denver, Colorado. Robert A. Wright, editor. 1978
18. *Energy Resource Recovery in Arid Lands.* Fort Lewis College, Durango, Colorado. K. D. Timmerhaus, editor. 1979
19. *Arid Land Plant Resources.* Texas Tech University, Lubbock, Texas. J. R. Goodin and D. Northington, editors. 1979
20. *Origin and Evolution of Deserts.* Las Vegas, Nevada, and Lubbock, Texas. Stephen G. Wells and Donald R. Haragan, editors. 1983
21. *Pattern and Process in Desert Ecosystems.* El Paso, Texas. W. G. Whitford, editor. 1986

Small Water Impoundments in Semi-Arid Regions

The Committee on Desert and Arid Zones Research of the Southwestern and Rocky Mountain Division of the American Association for the Advancement of Science Statement of Purpose

The objective of the Committee on Desert and Arid Zones Research is to encourage the study of phenomena relating to and affected by human occupation of arid and semiarid regions, primarily within the areas represented by the Southwestern and Rocky Mountain Division of the A.A.A.S. This goal involves educational and research activities, both fundamental and applied, that may further understanding and efficient use of our arid lands.

Mailing Address

Dr. M. Michelle Balcomb, Executive Officer
SWARM/AAAS
Colorado Mountain College
3000 County Road 114
Glenwood Springs, CO 81601

Small Water Impoundments in Semi-Arid Regions

Edited by

John L. Thames
and Charles D. Ziebell

University of New Mexico Press

Albuquerque

Library of Congress Cataloging in Publication Data

Small water impoundments in semi-arid regions.

(Contributions of the Committee on Desert and Arid Zones Research of the Southwestern and Rocky Mountain Division of the American Association for the Advancement of Science ; vol. 22)
Based on a symposium held in Lubbock, Texas in March and sponsored by the Committee on Desert and Arid Zone Research of the American Association for the Advancement of Science.
Bibliography: p.
Contents: Water resources of small impoundments in dry regions / Kenneth G. Renard—Limnologic features of reservoirs in arid lands / Gerald A. Cole—Pond limnology, with applications to arid-area ponds /

William J. Clark—[etc.]
1. Reservoirs—Arid regions. 2. Ponds—Arid regions.
3. Playas—Arid regions. 4. Water use—Arid regions.
5. Limnology—Arid regions. 6. Aquatic ecology.
I. Thames, John L. II. Ziebell, Charles D.
III. American Association for the Advancement of Science. Committee on Desert and Arid Zones Research. IV. Series: Contribution . . . of the Committee on Desert and Arid Zones Research ; no. 22.
TD395.S58 1988 551.48′2′09154 87-30132
ISBN 0-8263-1053-2

First edition.

Contents

Tables

Figures

Preface

This book evolved from the symposium "Small Water Impoundments in Semi-Arid Regions" held at Lubbock, Texas, in March 1984 and organized by the Committee on Desert and Arid Zone Research of the American Association for the Advancement of Science. The symposium focused on the physical and limnological characteristics of small water bodies in semi-arid environments as well as the ecological relationships important for the wise planning and development of dry-land water resources.

The water relationships of arid and semi-arid zones are perhaps more critical to a greater number of people than those of more humid regions. Dry regions cover about one-third of the earth's land surface; about one-half of this area is inhabited by a human population of nearly 700 million individuals. Population pressures combined with poor land use practices are turning potentially productive dry lands, already in fragile ecological balance, into unproductive deserts over most regions of the earth.

Paradoxically, arid and semi-arid lands are usually characterized by abundant sunlight and mineral rich soils; with water, they can become highly productive and support a much larger human population than they now do. The development and wise use of water resources, particularly the development and use of renewable surface water resources, can assist in increasing this productivity. Small water impoundments offer opportunities for this development not only for water supplies in remote areas, but also as dispersed sources for aquatic production, wildlife habitat, recreation, and aesthetics. The eight essays in this book briefly address some of the various hydrological and biological relationships of small water impoundments in dry regions, and they range from general concepts to specific studies important in the planning and use of small water impoundments.

Reservoirs and impoundments in dry regions are dramatically affected by hydrologic factors and events on upstream watersheds. These factors, including precipitation, runoff/water yield, soil profiles, and sediment yield, are discussed and illustrated in diagrams, maps, and graphs. A physically-based water resource simulation model (SPUR), designed to aid managers and researchers in decision making, is presented in Chapter 1. Equations and diagrams describe the five components (climate, hydrology, plants, animals, and economics) of the model.

Reservoirs in arid lands, whether small stock tanks or larger man-made lakes, are influenced by unique climatic, geologic, morphologic, and limnologic features. These features are examined and differences between small and large impoundments are underscored. Studies done in arid and semi-arid regions of the world are cited and results are discussed in Chapter 2.

Because ponds are highly variable, they defy generalities and predictions. However, such general characteristics as shallow depths and small volumes along with surface areas and watersheds are discussed. Results from a study of five ponds in Brazos County, Texas, (measuring temperature, dissolved oxygen, water chemistry, and biology) are shown in tables and graphs. In addition, features of arid-area ponds, most of which are man-made and unique, are identified in Chapter 3.

An important function of small impoundments in a multiple-use setting is providing fish habitats. Factors affecting fish communities in ponds in semi-arid regions are described. Specifically, species composition, the environmental effects of water level fluctuation, clarity, fertility, and pond management are considered. Studies done on small impoundments in Texas are cited in Chapter 4, and frequency of occurrence of fish species in south Texas ponds and west central Texas floodwater-retarding structures are listed in a table.

Elevated levels of pH associated with the development of various types of aquatic plants, including macrophytes and algae, have resulted in fish mortality in trout fisheries of the southwestern United States. Three lakes, differing morphometrically but typical of those of the mountains of Arizona, were studied between May and September. Levels of pH varying with the sizes and depths of the lakes, the amount of aquatic plants and algae present, and the time of day are presented graphically and tabularly. Macrophytic plants and algae are identified and the resulting elevated pH is explored. In addition, several procedures are suggested by the authors in Chapter 5 to minimize fish mortality in trout fisheries.

Desert playas, common in the southwestern United States, have not been intensively studied, despite the fact that they are simple systems that can be easily manipulated to test ecological theories. Data from a playa located in the Jornada Long-Term Ecological Research Area in New Mexico were compared with earlier data and those from other playas to identify the effects of flooding. The hydrologic, chemical, vegetative, and animal responses to flooding were studied and the results are shown in tables and graphs in Chapter 6. In addition, interspecific trophic interactions, decomposition, determination of heterotrophy, comparison of temporary ponds, physiological adaptations, dispersal, and research needs were discussed.

In the Southern High Plains of Texas, there are 18,000 playa lakes providing habitat for wildlife. Man's activities in this region can be both beneficial and detrimental to the playa environments. A study of seven pitted playas (four terraced and three control) designed for water storage measured use by ducks and other water birds; benthic invertebrates and residual seeds; swimming and plant-clinging invertebrates; and vegetation. Results are given in tables and graphs, and the configuration of the playas is illustrated by photographs in Chapter 7.

Because land use activities in arid environments require additional water and energy, planning decisions must be based on dependable information on the costs and benefits of providing these resources. The land use planner today is challenged to provide immediate solutions to problems and to contribute scientific information that will benefit society. Three land use projects in Arizona were designed to use previously unused water and to redistribute it to the environment. Tables, graphs, and diagrams are presented in Chapter 8 to illustrate the factors that were considered in planning projects to control erosion, provide recreational opportunities, benefit wildlife, control channel erosion, and dispose of effluent.

It is hoped that the essays in this book will stimulate further research, conservation, and development of the scarce water resources of arid and semi-arid zones.

1

Water Resources of Small Impoundments in Dry Regions

Kenneth G. Renard

SUMMARY

Reservoirs and impoundments in dry regions are dramatically affected by hydrologic factors and events on upstream watersheds. In the southwestern United States, extreme temporal and spatial variability in precipitation often results in erratic runoff/water yield patterns. Runoff in ephemeral streams is subjected to large infiltration losses in the channel alluvium and decreased peak discharge with increasing drainage area. As a result of the hot and dry conditions, there are high evapotranspiration rates and sparse vegetation. The limited vegetation, along with steep gradients in some areas, contributes to low infiltration and high erosion rates. Shallow soil profiles and high sediment yields also characterize arid areas. Ranching economics in these areas cannot justify land management and conservation efforts that require capital expenditures. Because many of these characteristics of arid regions are difficult to measure over extensive areas, water resource modeling has become important in planning small impoundments. One such model, SPUR (Simulation of Production and Utilization of Rangelands) considers climate, hydrology, plants, animals, and economics to aid resource managers and researchers in decision making.

Introduction

There is a lovely road that runs from Ixopo into the hills. These hills are grass-covered and rolling, and they are lovely beyond any singing of it. . . .

> The grass is rich and matted, you cannot see the soil. It holds the rain and the mist, and they seep into the ground, feeding the streams in every kloof. It is well-tended, and not many cattle feed upon it; not too many fires burn it, laying bare the soil. Stand unshod upon it, for the ground is holy, being even as it came from the Creator. Keep it, guard it, care for it, for it keeps men, guards men, cares for men. Destroy it and man is destroyed.
>
> Where you stand the grass is rich and matted you cannot see the soil. But the rich green hills break down. They fall to the valley below, and falling, change their nature. For they grow red and bare; they cannot hold the rain and mist, and the streams are dry in the kloofs. Too many cattle feed upon the grass, and too many fires have burned it. Stand shod upon it, for it is coarse and sharp, and the stones cut under the feet. It is not kept, guarded, or cared for, it no longer keeps men, guards men, cares for men. The tithoya does not cry here any more.
>
> The great red hills stand desolate and the earth has torn away like flesh. The lightning flashes over them, the clouds pour down upon them, the dead streams come to life, full of the red flood of the earth. Down in the valleys women scratch the soil that is left, and the maise hardly reaches the heights of a man. They are valleys of old men and women, or mothers and children. The men are away, the young men and girls are away. The soil cannot keep them anymore (Paton, 1948).

Although the quote from Paton's *Cry, the Beloved Country* was written about conditions in southern Africa, the condition is being repeated all too frequently worldwide. The quote seems especially appropriate to a consideration of small water impoundments. As ecologists, environmentalists, and/or conservationists, we are all aware that what happens on the watershed can have dramatic consequences on the dynamics of what happens in the reservoir or impoundment.

The hydrology/water resources of headwater basins in much of the southwestern United States are dominated by:

(1) Extreme variability in precipitation
  - (a) temporally and
  - (b) spatially.

(2) Runoff produced in ephemeral streams, which are dry most of the time,
  - (a) as runoff traverses the dry streambed, transmission losses reduce the flow;

(b) water yield per unit area decreases with increasing drainage area;
(c) peak discharge per unit area is large near the center of a thunderstorm.

(3) Arid and semiarid conditions, which predominate result in
(a) high potential evapotranspiration and
(b) sparse vegetation.

(4) Physiographic and soil conditions, which result in
(a) low infiltration and
(b) high erosion rates because of steep gradients and sparse vegetation.
(c) steep streams and large alluvial supplies leading to high sediment yield.

(5) Ranching economics that lead to
(a) minimal use of agricultural chemicals;
(b) minimal efforts for land modification;
(c) minimal incentive for conservation programs, and
(d) minimal incentive for range improvement/rehabilitation.

PRECIPITATION

Hershfield (1962) considered annual precipitation totals for a large number of gages around the conterminous U.S. and showed that the maximum coefficient of variation is in the Southwest, with the highest values in southern Arizona and southern California (Fig. 1.1). Thus, in addition to the aridity of this same area (except for high-elevation mountains), the highly variable annual rainfall can result in an even more erratic runoff/water yield pattern.

In southeastern Arizona, about ⅔ of the annual rainfall occurs as thunderstorms during the July-September period. These thunderstorms are typically of short duration (1 to 2 hours), high intensity (up to 10 in/hr or 250 mm/hr for 5 minutes is common), and limited areal extent. The intensity pattern within a given storm is also highly variable, as indicated in Figure 1.2 where 4 raingages on the Walnut Gulch Experimental Watershed, with large amounts in a short period, were compared. Hyetograph differences for these storms are even more dramatic when presented as dimensionless distribution graphs.

Practicing hydrologists are continually plagued in thunderstorm dominated runoff areas (such as the southwestern U.S.) with daily rainfall totals as the only data available for design. Thus, if a reservoir is to be designed and precipitation data are available for a location within a project, the hydrol-

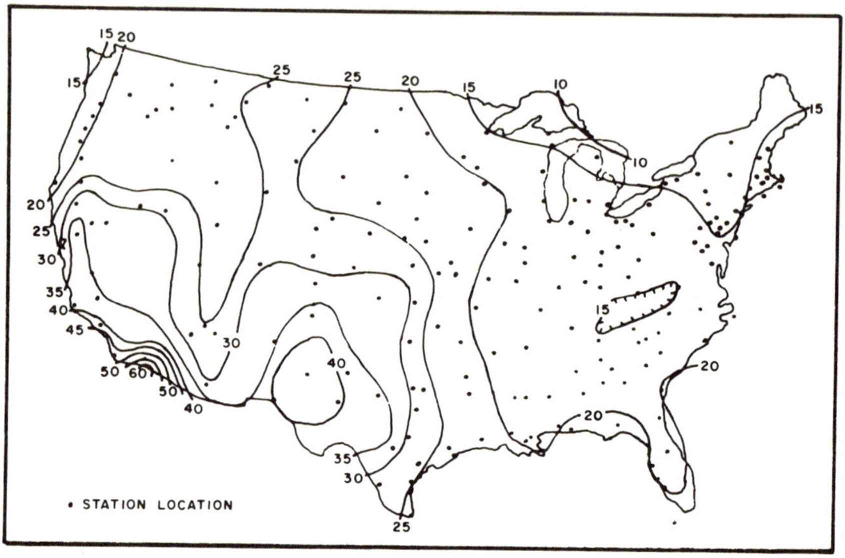

Figure 1.1. Coefficient of variation of annual precipitation in percent (Hershfield, 1962).

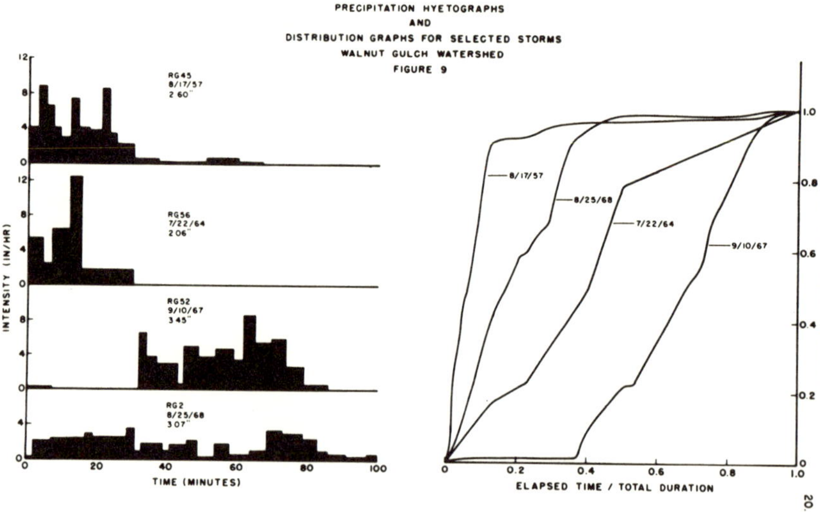

Figure 1.2. Precipitation hyetographs and distribution graphs for selected storms on the Walnut Gulch watershed (Renard, 1970).

ogist must decide how much runoff might be expected from the available rainfall record. Furthermore, to preclude the disaster of the impoundment being overtopped and destroyed, a spillway (principle and emergency) must be designed for some peak discharge and some total runoff volume. For example, if the rainfall record included a maximum of 3 inches (75 mm) on any day in the record, it would make a great deal of difference whether those 3 inches occurred in 1, 2, 6, or 24 hours, because 3 inches in 24 hours might not produce any runoff (depending on the soils and vegetation), whereas the same amount of rainfall in 1 hour might result in a very large runoff.

For this reason, hydrologists such as those at the Southwest Rangeland Watershed Research Center are developing probabilistic models which disaggregate daily rainfall amounts into a hyetograph so that precipitation excess (precipitation minus infiltration) can be computed with the use of a time-dependent infiltration function such as was developed by Rawls, Brakensiek, and Miller (1983).

Spatial precipitation variability associated with air mass thunderstorms represents still another design problem for the hydrologist. Unfortunately, the number of precipitation networks available for quantifying such variability is extremely limited (Osborn, Lane and Hundley, 1972; Hershfield, 1971).

One precipitation network is the U.S. Department of Agriculture's Walnut Gulch Experimental Watershed, in southeastern Arizona, which was established in 1961. The 57.7mi^2 drainage area (150 km^2) has a network of almost 100 recording precipitation gages on, or adjacent to, the study area. The network facilitates the definition of the rainfall on both individual storm events, as well as the aggregate from seasonal and annual totals. Some of this variability is indicated in Figure 1.3 for an event on July 16, 1967 and for the 1967 annual total. Data from networks such as these have been used with considerable success in modeling thunderstorm depth-area relationships (Renard and Brakensiek, 1976) which, in turn, are then used in water resource simulation models for varied climates in the continental U.S.

RUNOFF

In most dry regions, the water table is considerably below the stream channel invert, especially in the headwater areas. The result is that the streams are influent rather than effluent. As such, the stream channels contain water only during unusual precipitation/storm events, and then

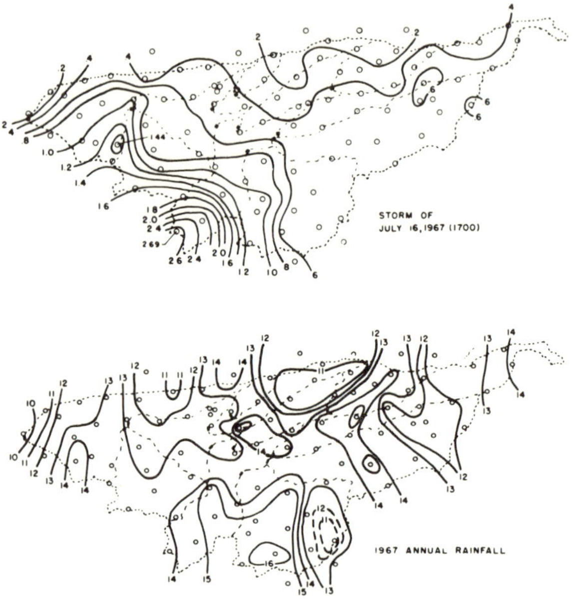

Figure 1.3. Isohyetal maps of the July 16, 1967 storm and the annual precipitation (in) for 1967 (Renard and Brakensiek, 1976).

generally for only short periods of time. Thus, the channels on a basin such as the Walnut Gulch Experimental Watershed are dry 99 percent of the time.

The isohyetal map for a typical air-mass thunderstorm with the hydrographs at consecutive streamflow stations on Walnut Gulch are shown in Figures 1.4a and 1.4b. Most of the heaviest part of the storm was concentrated on the 8.24 km^2 (2035 acre) subwatershed 11. Although there was some precipitation on the intervening area between subwatershed 11 and watershed 8, it was probably minimal relative to the amount from watershed 11. From flume 11 to flume 8, there are 6.6 km (4.1 miles) of channel, with a mean width of 11.6 m (38 feet). The 1000 cfs (cubic feet per second) (28.3 m^3/sec) peak flow with 49 acre-feet (AF)(60,400 m^3) of runoff measured at flume 11 was reduced by infiltration in the channel alluvium to slightly less than 700 cfs (19.8 m^3/sec) peak flow and 34 AF (46,800 m^3) of runoff at flume 8. For this flow, like many other storms on the upper portion of the watershed, most of the runoff was lost in the channels before

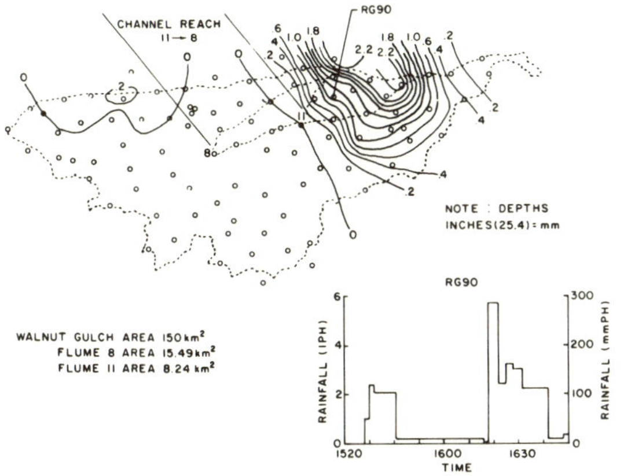

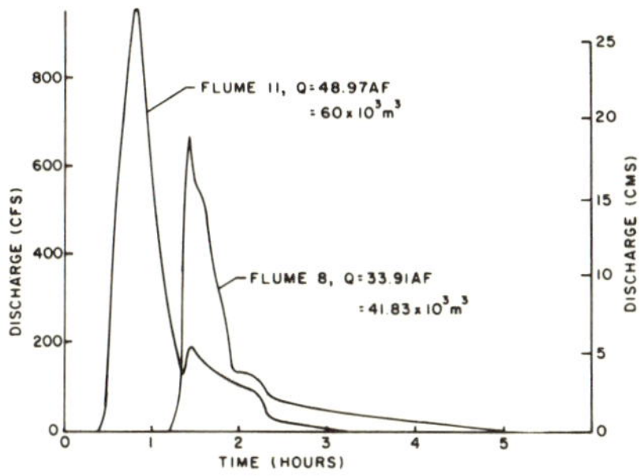

Figure 1.4a. Storm of July 30, 1966 centered on subwatershed 11. Isohyetal lines represent rainfall depths in 0.2 in increments.

Figure 1.4b. Observed hydrographs at subwatersheds 11 and 8 for the July 30, 1966 storm.

reaching the lowest station of Walnut Gulch. However, when storms occur near the outlet or when the channel alluvium is already wet, discharge at the outlet is greater.

Hydrologists unfamiliar with storm characteristics in the thunderstorm areas of the Southwest are usually surprised at the large peak discharges encountered in such streams. For example, on numerous occasions, peak discharges of 1500 cfs/mi^2 (16.3 m^3/sec/km^2) have been measured in those portions of the experimental area where a particularly intense thunderstorm was centered. Because of the spatial variability of the precipitation (leading to partial area runoff), and because of transmission loss, peak discharge per unit area decreases with increasing drainage area.

As with peak discharge, water yield generally decreases with increasing drainage area (Figure 1.5). By contrast, in more humid areas, water yield per unit area may increase with increasing drainage areas for example, Coshocton, Ohio in Figure 1.5. In other regions (Riesel, Texas), the water yield appears to be independent of drainage area.

The hydrologic balance of a watershed such as Walnut Gulch includes a rather large component for transmission losses (Figure 1.6). In fact, the hydrologic balance of Walnut Gulch contains almost 15 percent of the total water input (precipitation) as transmission losses, or a total of 6.7×10^6 m^3 for this 150 km^2 basin.

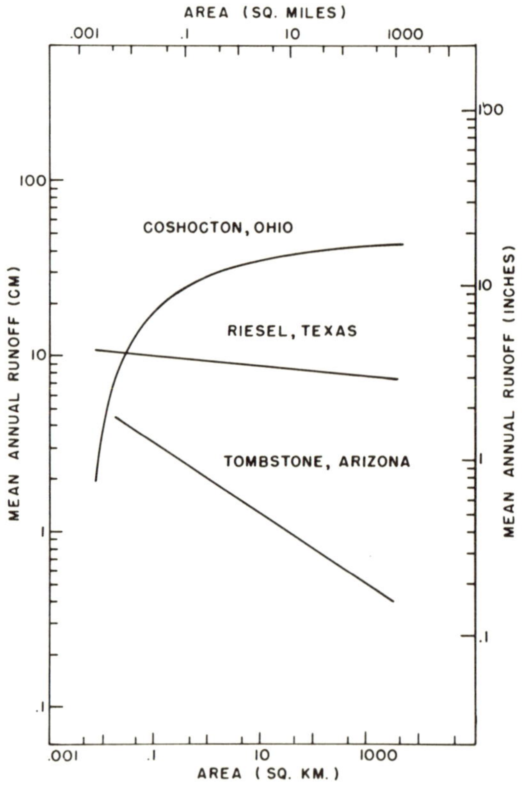

Figure 1.5. Mean annual runoff versus size of drainage area for several vicinities (Glymph and Holtan, 1969).

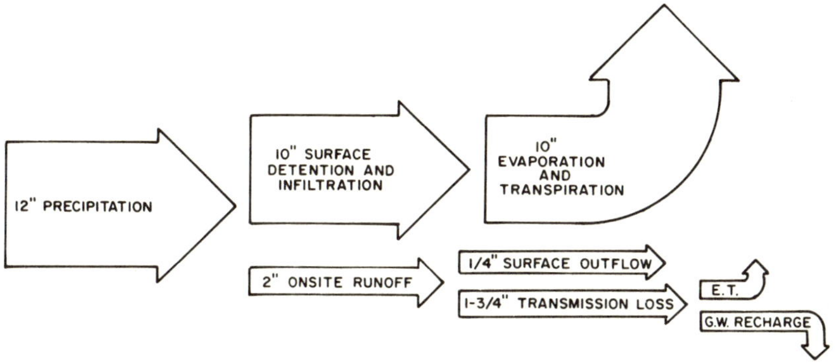

Figure 1.6. Water balance of Walnut Gulch Watershed (Renard, 1970) (1 in = 25.4 mm).

OTHER CHARACTERISTICS OF DRY REGIONS

Arid and semi-arid areas are characterized by moisture deficiencies, but more importantly, the potential for evapotranspiration is much greater than the annual or seasonal precipitation. For example, in southeastern Arizona, pan evaporation of more than 100 inches (2.54 m) is about 7 to 8 times the annual rainfall, enabling only vegetation capable of using the limited soil moisture rapidly and efficiently to grow.

In most dry regions, the watershed vegetation is composed of mixtures of plant communities (grasses, brush, forbs, cacti). Because of limited moisture, the percentage of the total land area covered by foliage of live plants is generally very low (on Walnut Gulch, for example, generally less than 10 percent, and often as low as 2 percent). Many of the bare areas between plants contain roots which affect the soil moisture beneath and between plants, and in turn, contribute to the soil structure-erosion problem. Most warm-season grasses are classified as C_4 plants, which are more efficient water users than the C_3 brush plants (most grass plants use about ⅓ the water of a brush plant per equal amount of biomass product). Thus, in soil moisture modeling, which is essential for runoff prediction, not only must the plant density be considered, but also the composition of the plants using the soil moisture.

PHYSIOGRAPHIC AND SOIL CONDITIONS

In most dry regions of the southwestern U.S., the topography is dominated by rolling hills or outwash from major mountain systems. Generally, land gradients are steep; slopes of over 50 percent are common.

Soils formations in most dry regions reflect the absence of moisture. In fact, because of the steep gradients, there is often an absence of an appreciable A horizon. A soil morphologist accustomed to conditions in more humid areas might be inclined to state that the western soils (except those of major irrigated valleys) are often really only partially weathered geologic material deposited as outwash from mountain systems.

The rate and amount of infiltration in dry-regions are often low. Low infiltration amounts are due to the short duration of the thunderstorms, shallow soil depth which limits the soil reservoir, the presence of impervious layers like caliche (Figure 1.7), and/or the presence of impeding layers at the soil surface due to such biological activity as algal crusts or cryptogams.

Erosion from dry regions is more of a problem than is generally thought. The amount of eroded material passing a given point on a stream system is known as sediment yield. In the sparsely vegetated areas of mixed brush-grass cover, where thunderstorms dominate and overland flow (Hortonian flow) is common, the sediment yield may well be at the maximum reported by Langbein and Schumm (1958) (Figure 1.8). The figure illustrates the importance of erosion/sedimentation processes in dry regions. Furthermore, these high sediment yield values help to explain the presence of shallow soil profiles, and often account for the absence of an A horizon. The most desirable type of soil, containing organic matter and nutrients, is often non-existent, because erosion removes it as fast as soil is developed from the parent geologic material. Erosion also explains the so-called erosion pavements (coarse material > 10 mm) dominating the soil surface in many sparsely vegetated areas (Figure 1.9).

In many larger streams, a large alluvial supply, high stream gradients (resulting in extraordinary velocities), and high peak runoff per unit area create large sediment yields. Two contrasting morphologic tendencies are apparent in an ephemeral stream channel: the channel profile tends (1) to be convex due to transmission losses in the normally dry channel alluvium, and (2) to be concave due to more flow downstream because of tributary inflow. Thus, the stream behavior is often difficult to predict. Sediment transport varies in ephemeral as well as perennial streams; in addition, the temporal and spatial variability in precipitation (where thunderstorms dominate runoff), combined with transmission losses, further complicate

Figure 1.7. A typical soil profile in the semi-arid areas of southeastern Arizona. The profile, which is often 70 percent gravel by volume, generally has a caliche layer at depths up to 1 m, which restricts moisture movement and root development.

the prediction of stream behavior. On most of the larger channels of Walnut Gulch, the channel profile is nearly constant, varying only slightly from the 1 percent mean value. Thus, detailed modeling of water yield, sediment yield and the transport of chemical pollutants from dry land watersheds is complex.

However, Renard and Laursen (1975) have developed a conceptual model to describe ephemeral stream phenomena mentioned above. It includes a stochastic runoff model (Diskin and Lane, 1976; Lane and Renard, 1972), the Manning equation, and the Laursen (1958) sediment transport

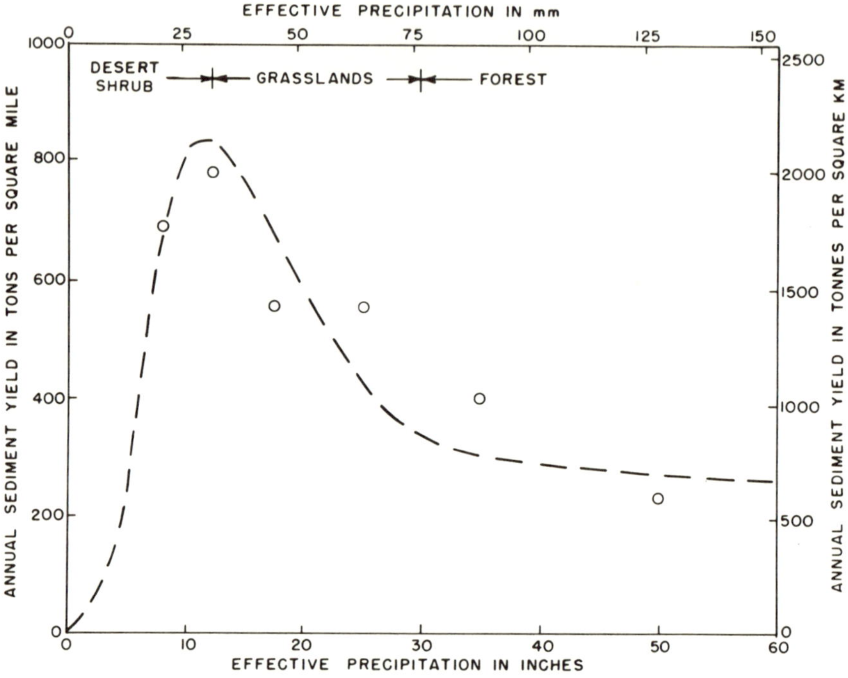

Figure 1.8. Effect of rainfall variation on sediment yield determined from records at sediment sampling stations (modified from Langbein and Schumm, 1958).

relation. The model was used to simulate the relationship between average runoff, sediment yield and drainage area (Figure 1.10). As expected with the large transmission losses, the model predicts decreasing sediment yield with increasing watershed area. To illustrate the sensitivity of the model to changing runoff, the bed material size was kept constant, and the generated runoff events increased and decreased for a 226-ha (560-ac) watershed (Figure 1.10). The variability in sediment yield corresponding to the runoff variability produced was appreciable, and larger than the runoff variability. To further illustrate the model's sensitivity, the mean grain size was altered in the model to reflect observed sediment-size distribution changes in the prototype channel. The resulting sediment yield variability is shown and labeled "range" for a 9,510-ha (23,500-ac) drainage area in the lower portion of the figure. Under prototypical conditions, the sediment transport probably adjusts in some selective process with particle

Figure 1.9. The erosion pavement in this photograph is the residual coarse material following erosion of the fine fractions from soil profile such as that in Figure 7.

shear changes associated with runoff variability. The bed composition at any time, therefore, responds to the rate at which material is being supplied to the channel from sediment sources during various hydrologic events and the rate at which material is being removed from a stream reach.

RANCHING ECONOMICS AND LAND MANAGEMENT

In the dry regions of the western United States, economics of the ranching industry are such that land management and conservation efforts requiring capital expenditures are often too costly to be undertaken. Similarly, because of the expense, the use of agricultural chemicals, such as nutrients for forage production, herbicides for brush control, and insecticides for insect control are generally minimal. Thus, nonpoint chemical pollution is generally not a problem in dry areas, although geologic weathering and natural chemical cycling does occur. There are, however,

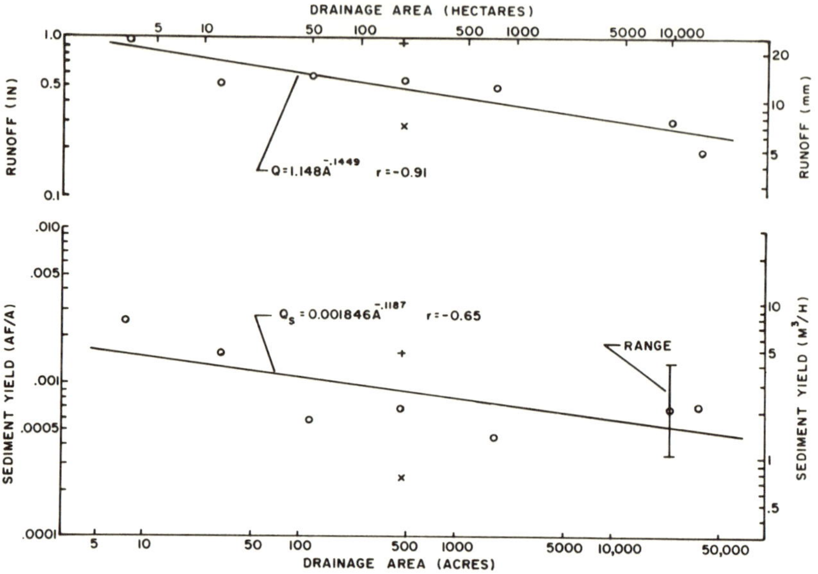

Figure 1.10. Average annual runoff and sediment yield as functions of drainage area for watersheds in southeastern Arizona based on Walnut Gulch simulation data. Equations are for English units (Renard, 1977).

frequent problems with biological contamination of streams and water impoundments.

The chemical aspects of water quality from rangelands have been investigated on Walnut Gulch by sampling at the watershed outlet and from three small interior subwatersheds with varied soils, vegetation, and land use. Schreiber and Renard (1978) observed high early-season nitrate concentrations (as high as 1 ppm) in runoff, which they felt reflected microbial activity in the soil and low plant uptake (plants still dormant). Phosphate relationships were different from those of nitrate, with each area contributing a relatively constant amount per event (one disturbed area had a consistently higher P concentration). The sum of the cations from all areas was relatively low (1 to 1.5 meq/l or milliequivalent per liter) with the highest values from the largest heterogenous area. The chemical constituents examined and reported in this work were within acceptable ranges for most uses, assuming prior removal of sediment found to be excessive in all water supplies.

Stephenson and Street (1978) found that typical rangeland cattle operations, such as are encountered in southwestern Idaho, will probably result in coliform bacterial pollution along various reaches of streams draining rangeland; however, these bacterial concentrations will vary according to the number of cattle and their access to streams, as well as according to physical and hydrologic characteristics of the flow regime and to climatic conditions. Continuous monitoring of streams under these conditions will probably show fecal coliform counts frequently in excess of water quality standards. This is even more likely for runoff samples taken in pastures used during winter months for feeding operations.

Stephenson and Street (1978) also reported that runoff from rainstorms increased both total and fecal coliform concentrations. The increase was observed on streams draining from summer grazing ranges with limited management and in streams adjacent to winter pastures; runoff from snowmelt had little effect. Total coliform counts varied more with change in streamflow than did fecal coliform counts. In fenced summer range allotments under deferred grazing management, the effects were the same, except that bacterial counts were not as high or persistent. The decrease in bacterial concentrations at several downstream sampling sites indicated that certain stream segments were self-purifying. The presence or absence of livestock along the streams overshadow any effect variations in chemical concentration of the water might have on bacterial concentrations.

In an extension of the earlier work, Stephenson and Rychert (1982) suggested that the elevated fecal coliform indicator counts observed in streams result primarily from a suspension of stream bottom sediment and organic matter, rather than from sources extraneous to the stream at the time of increased runoff. They suggest that the organic matter content of the sediment may have a critical influence on the survival and/or multiplication of the bacteria. Their results show *Escherichia coli* concentrations of bottom sediments to be from 2 to 760 times greater than from the overlying water. *Escherichia coli* concentrations of bottom sediment were found to be resuspended following disturbance simulation and a rainstorm event, contributing to pollution of the overlying waters. They suggested that microbial analysis of bottom sediments should be done as a part of any water-quality evaluation for rangeland streams. Depending upon the ultimate use of the small impoundment being evaluated, microbial analysis may be critical to the development plan.

Rangeland managers are frequently concerned with revegetation on deteriorated rangelands (Cox et al., 1983). Root plowing (pulling a knife beneath the soil surface to sever roots of woody plants) and reseeding were performed on a 42.5 ha (105-ac) subwatershed in Walnut Gulch in 1970 to measure the consequence of such a program on runoff and erosion/sediment yield. Although there was a period of several years following the treatment (Simanton, Osborn and Renard, 1977) where the runoff, as a function of precipitation, was ill-defined, the post-treatment and pre-treatment runoff characteristics are very similar. The sediment yield, following treatment (grass dominated), has been only about 1/7 of that for the pre-treatment (brush-dominated) area. However, the cattle grazing capacity of the improved condition, although about 10 times that for the brush, was still not sufficient to justify the treatment cost.

Although there have been numerous evaluations of grazing practice effects on infiltration (see Gifford and Hawkins, 1979), the results have often been inconclusive, or, at best, weakly defined regarding grazing and infiltration consequences. Thus, although research on range renovation efforts continues, currently the most viable alternative for range improvement/conservation is some system of grazing rotation. The method most prominent in the literature is the Savory system (Heitschmidt and Walker, 1983) developed in Africa. Like many other grazing efforts, this system needs to be evaluated in terms of effect on the soil-plant environment.

WATER RESOURCE MODELING FOR SMALL IMPOUNDMENTS

The development of water resource models has closely paralleled the development of digital computers. As recently as four decades ago, design work in hydrology was restricted to the use of nomographs and simple empirical equations. The current trend is toward elaborate digital computer programs capable of describing phenomena that are difficult to measure under field conditions.

Although many have recently written on modeling water resource systems (Branson et al., 1981; Haan, Johnson and Brakensiek, 1982; Dooge, 1973; Singh, 1982a, b, c, d), a brief overview of some of the techniques involved is in order. Figure 1.11 illustrates one way of conceptually classifying water resource models. In many instances, when a process (such as rainfall-runoff) is so complicated as to be ill-defined or when the variability involved is extreme, the modeler may resort to a stochastic model (shown diagramatically as a die). This type of model follows probabilistic laws, while the process develops sequentially in time, unlike purely probabilistic models that are not time-dependent. Mystery models (also called magic or black box models) utilize a given input to produce an output through a mathematical transformation or function (the function may or may not have physical significance). An analytical component model (often called causal model) uses the physics of the process(es) to produce an output. Although causal models are often the most expensive to develop and operate in a prediction mode (more algorithms involved, greater computer cost, and

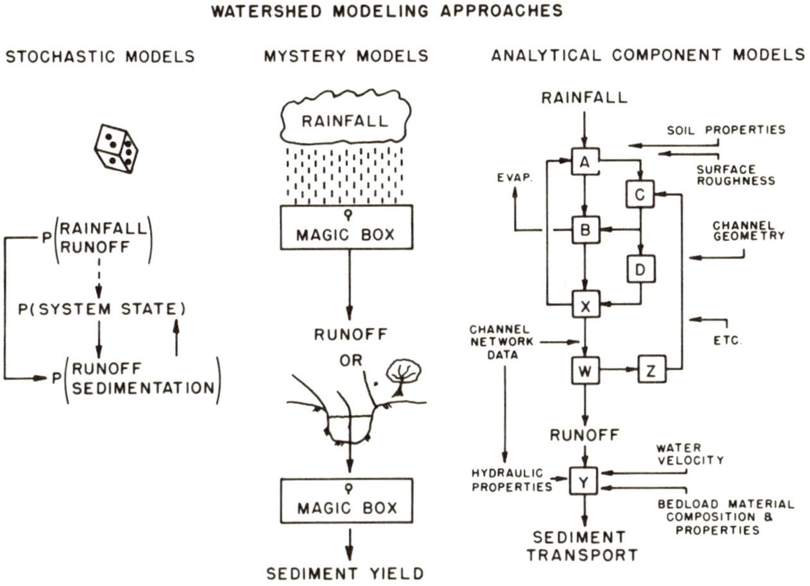

Figure 1.11. Approaches to sediment modeling from watersheds (Renard, 1977).

often more field data required), they generally provide more reliable results, especially when management alternatives are involved. In actual practice, a model for a specific problem may, in fact, include combinations of the three classifications.

A model developed recently by Agricultural Research Service (ARS) engineers and scientists for use on rangelands should be of special merit for a conference on "Small Water Impoundments in Dry Regions." The model, called SPUR (*S*imulation of *P*roduction and *U*tilization of *R*angelands) consists of five basic components: (1) climate, (2) hydrology, (3) plant, (4) animal, and (5) economics. Input data requirements and model outputs are described in a U.S. Department of Agriculture pubication edited by Wight (1983).

SPUR is a physically based rangeland simulation model designed to aid both resource managers and researchers. It can be applied to a wide range of conditions with a minimum of "tuning" or "fitting." It provides a basis for management decisions by predicting herbage yields, livestock production, runoff, and erosion. As a research tool, it helps identify research needs, improves the organization and transfer of information, and provides a focus for ARS range research programs. Individual model components for the

SPUR model were developed by specialists in each area. The individual components were merged under the leadership of the project coordinator, Dr. Ross Wight, in Boise, Idaho. A diagram of the model is given in Figure 1.12. Further development, testing, and documentation of the model are underway. The reader is referred to the SPUR manuscript for further details (Wight, 1983).

SUMMARY

Water resources in dry regions are, by definition, in short supply; thus, competition for them is often dramatic. We must learn to use these limited resources wisely.

The hydrology/water resources of the dry regions of the southwestern United States are characterized by:

(1) extreme temporal and spatial variability;
(2) ephemeral runoff, which often results in large infiltration losses in the channel alluvium;
(3) sparse vegetation and high potential evapotranspiration;
(4) low infiltration in shallow soils with high erosion and sediment yields;
(5) minimal chemical water pollution and minimal capital for conservation/range rehabilitation efforts.

Mathematical modeling offers an objective way to analyze the hetero-

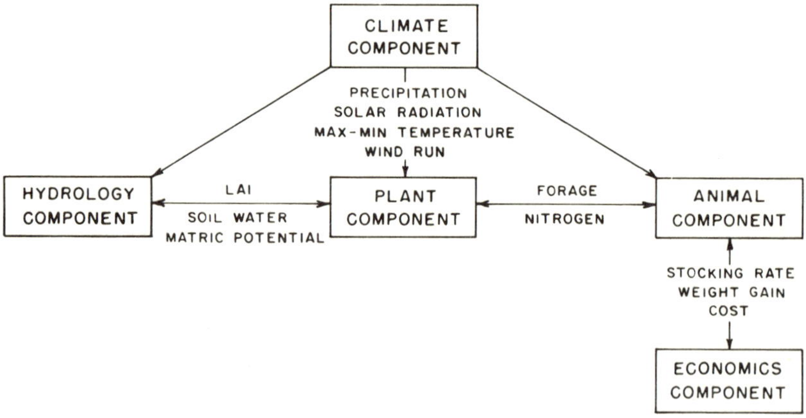

Figure 1.12. Interaction between various components of SPUR (Wight, 1983).

geneous conditions encountered in most dry regions. To better utilize the water resources of small impoundments, the designer must be able to assess the impact of various land use/management alternatives upon the water yield and water quality of the inflow to an impoundment. Sediment is the major water quality consideration for an impoundment, because unless there are major irrigated areas in the watershed contributing to the impoundments, only geologic chemical transport levels are likely to be encountered, and in general, these are low.

The SPUR (Simulation of *P*roduction and *U*tilization of *R*angelands) model, recently developed in the Agricultural Research Service of USDA, has been suggested as one model for predicting the inflow to a small impoundment.

REFERENCES

Branson, F. A., G. F. Gifford, K. G. Renard, and R. F. Hadley. 1981. *Rangeland Hydrology*. Range Sciences Series No. 1, 2nd Edition, Society for Range Management, 340 pp.

Cox, J. R., H. L. Morton, J. T. LaBaume, and K. G. Renard. 1983. "Reviving Arizona's Rangelands." *J. Soil and Water Cons.* 38(4):342–345.

Diskin, M. H., and L. J. Lane. 1976. "Application of a Double Triangle Unit Hydrograph to a Small Semiarid Watershed." *Hydrology and Water Resources in Arizona and the Southwest*. Am. Water Resour. Assoc., Arizona Acad. Sci., Hydrol. Sec., Tucson, Ariz., Vol. 6, p. 125–135.

Dooge, J. C. I. 1973. "Linear Theory of Hydrologic Systems." U.S. Dept. of Agric., Agr. Res. Service, Tech. Bull. 1468, 327 pp.

Gifford, G. F., and R. H. Hawkins. 1979. "Deterministic Hydrologic Modeling of Grazing System Impacts on Infiltration Rates." *Water Resources Bulletin* 15(4):924–934.

Glymph, L. M., and H. N. Holtan. 1969. "Land Treatment in Agricultural Watershed Hydrologic Research." In: *Effects of Watershed Changes on Streamflow*. Water Resources Symp. No. 2, Univ. of Texas Press, Austin, TX., p. 44–68.

Haan, C. T., H. P. Johnson, and D. L. Brakensiek. (eds.) 1982. "Hydrologic Modeling of Small Watersheds." Am. Soc. of Agric. Engrs. Monograph No. 5, 533 pp.

Hanson, C. L., E. L. Neff, and D. A. Woolhiser. 1975. "Hydrologic Aspects of Water Harvesting in the Northern Great Plains." Proc. Water Harvesting Symposium, USDA, ARS-W-22, p. 129–140.

Heitschmidt, R., and J. Walker. 1983. "Short Duration Grazing and the Savory Grazing Method in Perspective." *Rangelands* 5(4):147–150.

Hershfield, D. A. 1962. "A Note on the Variability of Annual Precipitation." *J. Appl. Meterology* 6(4):575–578.

Hershfield, D. A. (ed.) 1971. "Agricultural Research Services Precipitation Research Facilities and Related Studies." USDA-ARS 41–176, 118 pp.

Jackson, R. D., S. B. Idso, and R. J. Reginato. 1976. "Calculation of Evaporation Rates During the Transition for Energy-limiting to Soil-limiting Phases Using Albedo Data." *Water Resources Res.* 12(1):23–26.

Lane, L. J., and K. G. Renard. 1972. "Evaluation of a Basin-Wide Stochastic Model for Ephemeral Runoff from Semiarid Watersheds." *Trans. ASAE* 15(1):280–283.

Langbein, W. B., and S. A. Schumm. 1958. "Yield of Sediment in Relation to Mean Annual Precipitation." *Trans. AGU* 39:1076–1084.

Laursen, E. M. 1958. "The Total Sediment Load of Streams." *J. Hydr. Div., ASCE* 84(HY-1): 1530–1 to 36.

Osborn, H. B., L. J. Lane, and J. F. Hundley. 1972. "Optimum Gaging of Thunderstorm Rainfall in Southeastern Arizona." *Water Resources Res., AGU* 8(1):259–265.

Paton, A. 1948. *Cry, the Beloved Country.* Charles Schreibner's Sons, NY., p. 36.

Rawls, W. J., D. L. Brakensiek, and N. Miller. 1983. "Green-Ampt Infiltration Parameters from Soils Data." *J. Hydr. Div., ASCE* 109(HY1):62–70.

Renard, K. G. 1970. "The Hydrology of Semiarid Rangeland Watersheds." USDA-ARS 41-162. 26 pp.

Renard, K. G., and Laursen, E. M. 1975. "A Dynamic Behavior Model of an Ephemeral Stream." *J. Hydraulics Division, Proc. ASCE* 10(HY5):511–528.

Renard, K. G., and D. L. Brakensiek. 1976. "Precipitation on Intermountain Rangeland in the Western United States." Proc. Fifth Workshop, U.S./Australia Rangeland Panel, Boise, Idaho, June, 1975. Utah Water Research Laboratory, Logan, Utah, p. 39–59.

Renard, K. G. 1977. "Erosion Research and Mathematical Modeling." In: *Erosion: Research Techniques, Erodibility, and Sediment Delivery.* T. J. Toy (ed.) GEO Books, Norwich, England, p. 31–44.

Schreiber, H. A., and K. G. Renard. 1978. "Runoff Water Quality from Varying Land Uses in Southeastern Arizona." *J. Range Manage.* 31(4):274–279.

Simanton, J. R., H. B. Osborn, and K. G. Renard. 1977. "Effects of Brush to Grass Conversion on the Hydrology and Erosion of a Semiarid Southwestern Rangeland Watershed." *Hydrology and Water Resources in Arizona and the Southwest,* Office of Arid Land Studies, University of Arizona, Tucson 7:249–256.

Singh, V. P., editor. 1982a. "Rainfall-Runoff Relationships." *Water Resources Publication,* 582 pp.

———. 1982b. "Applied Modeling in Catchment Hydrology." *Water Resources Publication,* 562 pp.

———. 1982c. "Modeling Components of Hydrologic Cycle." *Water Resources Publication,* 590 pp.

———. 1982 d. "Statistical Analysis of Rainfall and Runoff." *Water Resources Publication,* 700 pp.

Stephenson, G. R., and R. C. Rychert. 1982. "Bottom Sediment: A Reservoir of *Escherichia coli* in Rangeland Systems." *J Range Manage.* 35(1):119–123.

Stephenson, G. R., and L. V. Street. 1978. "Bacterial Variations in Streams from a Southwest Idaho Rangeland Watershed." *J. Environ. Qual.* 7(1):150–156.

Wight, J. R. (ed.) 1983. "SPUR—Simulation of Production and Utilization of Rangelands: A Rangeland Model for Management and Research." *U.S. Dept. of Agric. Misc. Pub.* 1431, 120 pp.

2

Limnologic Features of Reservoirs in Arid Lands

Gerald A. Cole

SUMMARY

Reservoirs in arid lands, whether small stock tanks or large man-made lakes, are influenced by climatic, geologic, morphologic, and limnologic features. Water levels fluctuate, partly due to extreme variability in provenance (snowmelt, dammed rivers, or precipitation) and partly due to water losses from accelerated evaporation, greater snowpack sublimation, seepage through permeable materials in stream channels and in the impoundments themselves, reduced potential storage volume caused by erosion, and extraction for use by man. In arid climates, many reservoirs are warm, monomictic, and turbid, have strong density currents, and, sometimes, exhibit abrupt vertical temperature gradients. Organic materials transported by flash floods affect the chemical balance of ponds and lakes. Salinity may be higher (due to increased evaporation) or lower (where there are permeable substrates) than expected. The number of macrophytes present can be highly variable, as are the vertical oxygen profiles. Fauna of arid-land reservoirs may have low species diversity, probably due to low water fertility. However, species found in the impoundments may be widespread and endemic, with a plankton population typical of temporary waters. Impoundments can serve as stepping stones for the movement of species, while changes in water levels often interrupt successional trends.

INTRODUCTION

The reservoirs in arid and semi-arid regions throughout the world are diverse. In North America they range from the tiny earthen stock tanks (*represos*) of the arid Southwest or the large riverine lakes held by dams in steep-sided desert canyons to the very large impoundments on the Missouri River lying in a middle-latitude, steppe climatic zone. Between these extremes are the countless farm ponds of the Great Plains, with areas from 0.2–2 ha, and neighboring impoundments ranging up to 24 ha and with maximum depths of 2–6 m; these lie in natural depressions bordered by man-made dikes on one or more sides (Pennak 1963). Two other lake regions are considered here: the so-called Highland Lakes formed by damming the Colorado River of Texas, the largest of which is less than 10,000 ha; and those on the Trinity River of Texas, the most expansive of the most eastern of the arid-land lakes, having an area of only 4,412 ha. Admittedly, the latter are situated just outside the semi-arid region where they are subjected to a humid mesothermal climate.

A unique reservoir, found especially in places where water is scarce, is the modified natural lake. Certain lakes occupying basins formed by volcanic events have been altered with the aim of increasing storage volume. For example, dikes have raised the level of Kinnikinick and Ashurst lakes in Coconino County, Arizona. Also, a long-ago breached lava dam was repaired to restore White Horse Lake, 55 km to the west. Many of the numerous deflation basins holding the playa lakes of the Texas high plains are marred by many man-made pits and/or elongate trenches designed to enhance water availability (see Chapter 7). Silvey and Stanford (1978) estimated that 66–75% of the mean annual runoff in the American Southwest was impounded at the time they wrote. Furthermore, they underscored the importance of one type, the southwestern riverine impoundment, by citing data from Geraghty and Miller (1973): 1.8 million ha are covered by such reservoirs. Whatever the type of impounded water in an arid climate, it is important and worthy of more study.

Rzoska (1966) outlined differences between riverine impoundments and natural lakes. First, the deepest part of the former is at the dam (portrayed earlier by Neel 1963, Fig. 21.10). The longitudinal profile resembles half a natural lake in which the deepest part is often near the center. Second, there is a longitudinal distribution of sediments and biota, reminiscent of a stream rather than a standing body of water. The absence of a well-defined sublittoral zone, the extreme fluctuations in water level, and the rapid turnover or flow-through time of the water and its nutrients are further contrasts. Except for fluctuations in surface level, the differences between natural and man-made bodies of water are less pronounced in small impoundments, and Rzoska's contrasts are not always applicable.

Margalef (1975, 184) considered the one hundred Spanish reservoirs he studied to "be typically stressed ecosystems" with respect to the manner by which nutrients are introduced and move rapidly through them. The ultimate stress is complete desiccation, an occasional occurrence in the life of a small pond in an arid climate. Silvey and Stanford (1978, 12) conlude that, in an area of extreme environmental gradients, reservoirs "are altered systems that never approach an ecological steady-state." This applies especially to impoundments in arid and semi-arid zones such as the American Southwest.

An early survey of reservoirs in North America focused on the impacts they have, although there were a few limnologic comparisons with natural lakes (Neel 1963). Lake Mead, bounded by desert shores of Arizona and Nevada, was discussed, but little emphasis was placed on differences between reservoirs in arid and humid regions. In a companion chapter, Dendy (1963) concentrated on farm ponds, emphasizing them as fish producers.

One goal of this chapter is to examine various features of reservoirs in deserts and semi-arid regions. A second goal is to underscore the differences between the small artificial pond and the larger man-made lake. The words of Silvey and Stanford (1978, 3) are relevant here: "Perhaps the most evident and perplexing limnological similarity among Southwestern reservoirs is really one of dissimilarity." That may be the best generality in this chapter.

It is not the aim of this chapter to treat the impact of a riverine impoundment on the effluent stream. There is, however, one effect that has biologic interest, although it has been overlooked to a great extent. A concrete dam built on a permeable surface serves as an example of a steady-state groundwater flow pattern; if the hydraulic head is lowered, a transient flow pattern results (see Freeze and Cherry 1979, Fig. 217a, b). The interstitial habitat, hardly explored in North America, is then enhanced along the banks of the tailwater stream. In the semi-arid West some remarkable faunal finds have been reported from interstitial waters in Colorado (Pennak 1968, 1971; Ward 1977), and others are anticipated.

PERTINENT CLIMATIC AND GEOLOGIC FACTORS

Reservoirs in arid and semi-arid climates are subjected to some natural phenomena that are not found in humid regions. Obviously, the combined effects of precipitation and temperature accelerate evaporation rates. For example, the net evaporative loss in an area where the mean annual precipitation and temperature are 25 cm and 15° C, respectively, would be about 125 cm per year, although what occurs at the lake surface might differ from this theoretical figure. There is the well-known story of how Lake Mead's evaporative loss was lessened after it began to receive cold water from the depths of Lake Powell. According to Hoffman and Jonez (1973) the evaporative loss in 1953 was 217 cm; during 1965–1969, after the closure of Glen Canyon Dam, the mean annual loss was 191 cm. The 26-cm difference represents a saving of almost 172 x 10^6 m^3 each year. The effects of intensified evaporation are better included in discussions of the physical limnology of arid-land reservoirs, but some other results will be mentioned here.

Theoretically, the absence of trees in prairie terrain and their relative scarcity in other dry regions reduces interception of precipitation and allows more to join the groundwater before evaporating. Where western coniferous forests prevail, the areas destroyed by fire are colonized rapidly by aspen (*Populus tremuloides*). Stands of these deciduous trees persist for many years before the climax conifers replace them. Thus, during the winter, snow can reach the ground in the aspen patches of an otherwise dense forest canopy. A snowpack on the ground in a dry climate, however, is subject to a significant loss by sublimation, lessening the gain brought about by scanty forest cover.

Water in stream channels is subject to loss by evaporation and seepage before it reaches the impoundment; in ephemeral streams the losses are often very high. In canyons of the West, dry except during rare periods of sometimes violent precipitation, there are tectonic fractures and step faults, diatremes, desiccation joints in lava, permeable porphyritic rocks, and porous sandstones. An example is the Woods Canyon drainage basin in Arizona (Scholtz 1968), where water loss by seepage is great. These are arid-land features that must be considered in the development of water impoundments for irrigation, water supply, and water conservation designs.

When the water reaches a man-made reservoir it may be subject to further loss by seepage. In the giant Lake Nasser, Egypt, for example, the basin is composed of both impermeable granite outcrops and permeable Nubian sandstone; the annual loss is from 0.6 to 2.0 km^3 through the sandstone, depending upon the lake level and the hydraulic head. Similarly, the porosity of the Navajo and Kayenta sandstones that bound Lake Powell are sources of seepage loss. In the small (26 ha) Upper Lake Mary, 16 km southeast of Flagstaff, Arizona, annual losses from evaporation and seepage are great. The lake, an impoundment on Walnut Creek, serves Flagstaff as a source of water; it lies over a N-S trending "graben fault" that accounts for much of the seepage loss. An unexpected loss by seepage from Lago del Oro, near Tucson, Arizona, was described by de Jong (1973).

The lake (113.3 ha) was impounded by an earthen dam in the channel of a boulder-strewn ephemeral stream. The annual loss by evaporation, 2.44 m, was about one-fourth of the lake volume, but even without that loss the highest rate of seepage noted would have drained Lago del Oro in less than five months.

Loss of potential storage volume is another characteristic of arid-zone reservoirs. Crumbling of newly flooded cliffs following dissolution of long-dry minerals, slumping of talus slopes, and the opening of fracture lines are phenomena that are far less common in humid climates. Tremendous rock spalls (50,000 m^3) have formed and plunged into the rising waters of Lake Powell (Potter and Pattison 1976). Thus, coarse rubble decreases the volume of some desert lakes, but fine-grained silting probably is more widespread and better known. In general, the sparse vegetation of desert regions has been subjected to overgrazing which has led to extreme sheet and gully erosion. Early estimates (Eakin and Brown 1939) revealed a mean annual capacity reduction of 3.8% in small Zuni Reservoir, New Mexico, in a region where overgrazing was permitted. Similarly, the authors calculated an annual volume loss of from 0.44 to 0.88% in the riverine Elephant Butte Reservoir in the same state. Later, Jester (1971) reported a 21.6% decrease in the capacity of the latter since it was impounded in 1916. Off-road driving, intensive grazing, and brush fires also increase erosion and lead to silting of reservoirs. In the Southern Great Plains, agricultural development and grazing are important factors. The loss of 44% of the volume in Lake Waco, Texas, during the first 17 years of its existence attests to this (Lind 1971).

ORIGIN OF RESERVOIR WATERS

Extreme variability in provenance of water is a characteristic of the arid-land reservoir. In some riverine impoundments the vernal floods brought about by the melting of the snowpack in distant mountains contrasts with low flows at other seasons. In small ponds of the desert and prairie, irregularity is a paramount environmental factor. Spatial and temporal variation, especially in precipitation and wind, weaken any generality made concerning the status of a single pond at any particular time. Thus, the small pond of arid zones stands in contrast to a larger lake in the same area and to a similar-sized pond in a humid region.

Most reservoirs larger than about 300 ha are riverine impoundments. In endorheic areas they are formed by damming large rivers that arise in less arid, exorheic regions and that flow through the desert on their way to the sea. Smaller ponds are situated on ephemeral streams or are located strategically in natural drainage patterns, relying on intermittent runoff for their water. Some hardly have a catchment basin, receiving most of their water directly from precipitation. The many reservoirs of the eastern Colorado Plains are supplied by groundwater pumped by windmills or by man-made ditches carrying water from small streams (Pennak 1963).

Rarely, small irrigation or stock ponds are constructed by damming the outflow of springs. An artificial limnocrene at Monkey Springs in southern Arizona is typical. In the larger river canyons, small tributaries arising from springs and seeps enter the main stream, contributing waters of diverse ionic proportions. With impoundment, these eventually are flooded to form an elongated lake with dendritic outline.

RESERVOIR MORPHOLOGY

Reservoir morphology differs from that of natural lakes in some respects, but generalizations do not always hold up. If the deepest water is located within a small area directly behind the dam, an impoundment would have a low volume development index. In three very large reservoirs, Kariba, Bhakra, and Nasser, the volume development indices are 0.73, 0.86, and 0.93, respectively. Similarly, the natural Mountain Lake, Virginia, impounded by a rock slide, has an index of 0.93. In contrast, the mean index of 675 large natural lakes is about 1.30, with a range of from 0.87 for Lake Michigan to 2.07 for Lake Winnipeg.

Riverine impoundments are characterized by exceptional shoreline development indices. Long, narrow fjord lakes and some others occupying elongate grabens or unusual glacial depressions may have indices greater than 5.0, but probably no natural lakes can approach values of 27.3 (Oahe, Missouri River) or 28.2 (Nasser, on the Nile). In a much different climate, the long irregular lakes formed by impounding the Tennessee River in Alabama, Tennessee, and Kentucky, for example, are no less spectacular. The shoreline development index does not separate impoundments of humid and arid regions. The southwestern represo and the farm pond of a humid zone usually are more nearly alike with respect to their low shoreline development indices than they are to neighboring elongate riverine impoundments.

Sometimes the main difference between large reservoirs of different climatic regimes is the mean depth. This applies especially to the large riverine impoundments and is due to the deep narrowness of western river canyons. Also, dams on the large Missouri River in the Great Plains physiographic province have created fairly deep lakes: from Fort Peck to Gavins Point the six reservoirs have a mean average depth of 18.7 m. The four reservoirs on the Trinity River in Texas are, on the average, 5.5 m deep (Harris and Silvey 1940). Farther west, the Highland Lakes have an average mean depth of 8.4 m; one of them, Travis Lake, has a mean depth of 20.5 m. The four lakes on the Salt River in Arizona have mean depths from 17 to 29 m. Small ponds in any climatic setting are shallow, and the above does not apply to them.

SOME PHYSICAL LIMNOLOGY

Annual Circulation Patterns

Most of the deeper southwestern reservoirs can be classified as warm monomictic, if they lie south of 40°N. In some years, though rarely, the Texas reservoirs Lake Worth and Lake Dallas on the Trinity River at about 220 m elevation and 33°N, develop an ice cover and can be categorized as dimictic in those years. Elevation, of course, affects the annual cycle: Clayton Lake, New Mexico, at 1,570 m elevation, regularly freezes although situated at 36°30′N. Some small Arizona lakes at high altitudes—Woods Canyon Lake at 2,294 m and 34°18′N, for example—have winter ice cover and are dimictic. Others, such as Kinnikinick and Ashurst lakes, occupy natural depressions that have been modified to raise water levels; they are not riverine or canyon impoundments. They show some fleeting summer stratification, but usually circulate except when a regular winter ice cover prevails. They are cold monomictic or warm thereimictic, the latter category having been proposed by Bayly and Williams (1973). Many shallow reservoirs farther north must also belong to this class.

In shallow southwestern ponds, both natural and artificial, intense solar radiation warms upper strata, effecting distinct density differences and a diurnal stratification. At night the superficial layers are cooled by back radiation and evaporation, setting up convection currents that destroy the stratification.

Evaporation affects more than ideal patterns of mixing. It can be a factor of consequence in annual heat budgets. Many years ago it was known that, in both natural (Neumann 1953) and man-made lakes (Anderson and Pritchard 1951) in arid regions, evaporative cooling sometimes overtakes the effect of solar warming; the mean temperature of the body of water begins to decrease before the heating season has come to a close. This does not occur in humid areas. This cooling phenomenon, described by Anderson and Pritchard (1951) in Lake Mead, must be far less pronounced now, if it exists at all, because the release from Glen Canyon dam of cool water from the hypolimnion of Lake Powell has lowered Mead's evaporation rates.

Turbidity and Light

Small southwestern ponds are usually turbid, with low Secchi-disc (Z_{SD}) transparency ($Z_{SD} < 10$ cm) and usually have a high euphotic zone; large reservoirs also are noted for their turbidity at times. Following flooding of the influent Nile and tributaries, the Z_{SD} of Lake Nasser is reduced to 20–25 cm. During high stages, the Little Colorado river of Arizona may reduce the transmission of light to less than 0.001% below its confluence with the Colorado River. This opacity persists in the Colorado and into the upper reaches of Lake Mead (Kubly and Cole 1979). Such turbidity is largely a phenomenon of arid zones. One recalls that the terms *anemotrophy* and *argillotrophy* came from studies of waters in dry regions (Hutchinson, Pickford, and Schuurman 1932; Hutchinson 1937).

The represo Little Triste near Chandler, Arizona, is argillotrophic with a mean Secchi-disc transparency of 6 cm; its coefficient of light absorption usually is about 19 (Barry 1980). Two lined ponds in Phoenix investigated by Foster (1973) held groundwater charged with tertiary-treated sewage effluent. Their mean total-dissolved-solids density was about 1.3 g/l, and they were remarkably rich in suspended algae, although much of the particulate matter was allochthonous (mean seston dry mass ca. 42 g/m^3; mean chlorophyll a, 183 mg/m^3). The coefficient of light absorption was commonly 4.0, although it ranged up to 10 on occasion.

Where reduced light penetration prevails, as in many small artificial ponds, a new dimension must be introduced in certain limnologic procedures. For example, when Foster (1973) attempted to estabish an assimilation number to relate chlorophyll density to the amount of carbon fixed per hour, she found unacceptable variance. When the reciprocal of the coefficient of light absorption was included in the equation, the standard error of the mean was reduced. The lined ponds clearly were autotrophic (P/R = 1.9). The above would not apply to the heterotrophic Little Triste, where sestonic chlorophyll hardly was demonstrable.

Temperature Profiles

Abrupt vertical temperature gradients are common in arid-land bodies of water. Usually this phenomenon is associated with, but not restricted to, salinity layering. Where turbidity caused by particulate matter is great, the absorption of heat in upper strata and the high coefficient of light absorption make for sharp temperature profiles. Typically, Barry's stock tank and

Little Triste had abrupt temperature profiles, and even Foster's two lined ponds in Phoenix at times revealed gradients of 0.65° C/cm in the upper 60-cm stratum. Larger opaque reservoirs are subjected to greater wind action and, therefore, do not exhibit such temperature clines except briefly during calm periods.

Density Currents

Density currents occur in riverine impoundments no matter what the climate. The TVA reservoirs (Gerking 1963), Elephant Butte reservoir studied by Bliss (see Grover and Howard 1938), and Lake Mead (Grover and Howard 1938; Hoffman and Jonez 1973) have been studied in detail with respect to this phenomenon. The classic example of a density current, however, comes from Lake Geneva in Europe, where the Rhone River enters. Strong density currents are not restricted to man-made lakes.

The density current has been considered a discrete entity within a lake, perhaps rich in nutrients denied to phototrophs in the upper illuminated stratum, and passing through the reservoir unaltered. This is part of the stress imposed on the reservoir ecosystem (Margalef 1975). Fisher and Smith (1983), however, showed that the density current is not so indestructible as had been believed, and they demonstrated the importance of making nearly continuous diel observations. They followed a plunging density current in Las Vegas Bay, Lake Mead, (an inflow with almost three times the salinity of the lake water) during two 14-day periods. Their observations and experiments (with dye) revealed that internal waves, and short-term formation and destruction of weak temporary thermoclines almost immediately sent 10% of the nutrients in the current to the lake surface. It should be mentioned that eleven years earlier the data of Koenig, Tew, and Deacon (1972, 112) who were studying phytoplankton in Las Vegas Bay, prompted them to note that the "density current was interrupted at intervals by dynamic events in the lake."

SOME CHEMICAL LIMNOLOGY

Early Nutrients and Oxygen Phenomena

Stock tanks in arid regions inherit an abundant nutrient supply from the animals that congregate along their shores. When violent, often localized showers occur, dung is swept by surface runoff into the ponds. In cases where the adjacent terrain permits the construction of a larger pond and the old stock tank and its peripheral catchment basin are flooded, the supply of nutrients is almost instant. The rise in potassium in the water of Little Triste following heavy rains has been attributed to the washing in of dry cow dung and rabbit pellets (Cole 1983). This might be an oversimplification. We still have much to learn about the ions carried by rain dripping from different species of desert shrubs.

Vertical oxygen profiles in deep reservoirs appear abnormal because so often there are negative heterograde curves associated with density currents; some are more enigmatic. Moreover, the distribution of water masses near the dam is affected by wind-generated currents meeting the nearly perpendicular dam surface. Analyses of a longitudinal profile near the mid-channel section of a riverine impoundment shows bizarre oxygen patterns as figured by Ellis (1940) for Elephant Butte. Back pumping, associated with pump storage activity, modifies the distribution of water masses and oxygen stratification even more. A striking and puzzling metalimnetic O_2 minimum in Lake Mead has existed since the reservoir was formed (Hoffman and Jonez, 1973; Paulson, Baker, and Deacon, 1980). The latter authors attributed the low point to in situ consumption of oxygen, either chemical or biological. Dense assemblages of *Dorosoma petenense*, threadfin shad, have been invoked by some workers to explain the oxygen minimum, but the threadfin was not introduced until 1953. The negative heterograde oxygen phenomenon termed the "D.O. dip" by Wiebe (1938, and subsequent papers) was associated with density currents in Norris Reservoir, Tennessee. The work of Dendy (1945) showed that the vertical distribution of fish in Norris did not explain the oxygen minimum.

Factors Affecting Salinity

A riverine lake in climatic areas where only natural closed basins exist is not closed if it has outlets, with the result that it does not become markedly saline. Sometimes a closed basin formed by impoundment in arid zones remains fresh because of a slowly permeable substratum. A few impoundments with permeable substrates such as those located on volcanic ash around Flagstaff, Arizona, are not as closed as they appear. Being filled only by direct precipitation and runoff in an area where the annual rainfall rarely attains 58 cm, they remain dilute because of flow-through due to seepage rather than to an effluent stream. Conversely, in some arid climates, marked salinity is owed to the substratum. High salinities are found in springs and pools of Chihuahua, Texas, and New Mexico, where underlying deposits were left by Permian seas.

Some small reservoirs supplied by intermittent streams become increasingly saline as evaporation proceeds. Dilution occurs when there is flow and the salinity level undulates inversely with the pond level. In areas outside of the region of closed basins, increasing salinity hardly exists. For example, the impoundments on the Trinity River of Texas (Harris and Silvey 1940) lie in the least arid region of all western reservoirs discussed here. There are probably no closed basins in areas receiving 62 cm of precipitation annually, the Fort Worth–Dallas mean, or where the annual net evaporation is less than about 50 cm.

Unusual Organic Compounds

In smaller impoundments in arid regions, introduction of organic material is unique. Unleached oak leaves, for example, driven to the lake by infrequent flash floods, carry phenolics in amounts that might be an environmental hazard. McConnell (1968) reported a flash flood import of unleached oak litter to Pena Blanca Lake, Arizona, in the amount of 750 g/m^2 of lake surface. Following this, the temporary phenolic concentration rose to 20 mg/l. Cold-water extracts of the detritus from *Quercus emoryi* (common in the watershed of Pena Blanca) analyzed for phenolics contained about 15 compounds, 85% tannins; the remainder were monomeric (non-tannin) phenolics (McConnell and Longo 1971).

In regions where precipitation is low and conifers are present, unleached needles are washed to lakes in flooding. These introduce the phenolic flavonoid taxifolin, which is similar to the oak phenolic quercetin. The former inhibits reproduction and lowers survival rates of *Daphnia pulex*. McConnell and Adams (1973), however, found no mortality in the test animals, *Helisoma, Daphnia,* and *Notropis,* exposed to phenolic compounds equivalent to about 20 mg/l, the Pena Blanca level. Whatever its effect, unleached (and this word should be stressed) allochthonous debris is a problem to the chemical balance of arid-zone ponds and lakes.

BENTHIC SUBSTRATA, ALGAL PERIPHYTON, AND LITTORAL MACROPHYTES

Flooding of western river canyons in many instances includes submergence of riparian trees. In other cases, only small shrubs are inundated and some species are remarkably resistant to decay. These shrubs serve as a substratum for periphytonic algae and animals. The ingenious scheme of Potter and Louderbough (1977), who used plastic Christmas trees of known surface area for sampling diatoms and macroinvertebrates, closely mimicked the natural situation. They used the Shannon-Weaver equation to calculate diversity indices ($H' \log_e$) of diatoms collected with the Christmas tree samplers. Their work was done in Lake Powell and the shrubshaped artificial substratum represented a change from the common use of glass slides.

In the shallows of reservoirs formed from impounding rivers in rocky canyons, macroscopic plants are scarce. There the littoral zone is defined by the lakeward limit of epilithic algae and the chlorophyllous periphyton on submerged terrestrial vegetation. Fluctuating water levels and inadequate substrata play the major roles in excluding littoral macrophytes. Sometimes at the upper end of an elongated riverine impoundment large quantities of sediment collect. A river delta forms and various pondweeds become established there, at a time when lack of proper substratum prohibits their colonization farther down the lake. Silvey and Stanford (1978, Fig. 3) show the development of a marshy habitat in the shallows of Lake Texoma where it receives the Red River and an alluvial fan has formed.

The four shallow reservoirs of the Trinity River basin in Texas, located in the most humid region of the arid-land lakes discussed here, support many macroscopic hydrophytes (Harris and Silvey 1940). The numbers of plant species in each reservoir seem to be related directly to the years since it was first impounded and inversely to the mean depth. When we turn to the smallest artificial ponds, we see another example of the difficulty in generalizing about them. The opacity of some ponds prohibits the establishment of a hydrophyte flora. In less turbid pools rooted pondweeds flourish, covering most of the bottom and contributing a high percentage of the pond's primary production.

ANIMALS OF THE ARID-LAND RESERVOIR

Diversity

Low diversity may be typical of animal populations in new and unstable reservoirs, although higher in others (see also Chapter 7). The diversity of the littoral invertebrates captured on the plastic Christmas tree substratum in Lake Powell was low, especially the equitability component. Only 28 species were found, and 90% of all the animals collected were species of the chironomid *Dicrotendipes*. The collecting was done 10 years after the Glen Canyon dam was closed (Potter and Louderbough 1977). The authors did not calculate diversity indices for their Christmas tree catches, but during one four-week period they collected 22,366 specimens, representing only 23 taxa.

Cosmopolitan Species and Some Others

Artificial aquatic habitats such as aquaria, canals, garden pools, quarry pits, and the range of from small to large impoundments are host to certain nearly cosmopolitan forms. Some of them seem to be fugitive species in the sense of Hutchinson (1951). Reservoirs in arid climates are not different in this respect. More than 20 years ago the freshwater medusa *Craspedacusta sowerbyi* was reported in Lake Mead by Deacon and Haskell (1963). Later, the same species was discovered by Kynard and Tash (1974) in Lake Patagonia, a small (107.7 ha) warm-monomictic lake in southern Arizona. It was found five years after a small creek was impounded to form the lake. *Craspedacusta* also occurs in quiet bays of natural lakes and has been found in metalimnetic strata of some natural glacial lakes when it was absent from surface waters (Dodson and Cooper 1983). It may be more widespread than has been suspected, and we have much to learn about it.

A second species (that coincidentally shares the same specific name) is the interesting gilled tubificid worm *Branchiura sowerbyi.* It occurs on all continents except Antarctica. It first appeared in the American West in an Oklahoma stock-watering tank (Evans 1959), and later in a canal in Tempe, Arizona, and in Saguaro Lake, an impoundment on the Salt River, Arizona, in 1963 and 1964, respectively (Cole 1966). Probably both the jellyfish and the worm are far more common than literature records would indicate.

The list of organisms that invade reservoirs includes some that are unwanted. The bivalves *Dreissena* in Europe and the Asian *Corbicula* in North America typify this phenomenon. Arid and humid regions do not differ in this respect. In Russian reservoirs, and especially those located in arid zones, the mass development of sponges was noted by Rzoska (1966). In the southwestern United States, large colonies of sponges were attached to rocks near the shore in Saguaro Lake, Arizona, during the 1950s and 1960s and perhaps since then.

Temporary-Pond Fauna

Small impounded bodies of water in the arid West develop a plankton population reminiscent of early stages in larger reservoirs. They, however, often contain species that are considered typical of temporary waters. Probably this is a function of astatic water levels, marginal areas being ephemeral in nature. The notostracan *Triops* and various anostracans are common in small, permanent stock tanks. Even the larger (146 ha) Ganado Lake in the Arizona portion of the Navajo Reservation supports a species of *Branchinecta* (Anostraca) along with fauna typical of permanent waters except for the absence of fish. Its level fluctuates widely; during the month of October 1964 the volume decreased about 1.4×10^6 m^3 (Cantwell 1965).

A further explanation for the occurrence of temporary-pond fauna in small impoundments of arid areas may lie in specific differences. The species found in vernal snowmelt ponds occupy a temporally predictable habitat. Belk (1977b) showed that, by contrast, the fairy shrimps occurring in unpredictable, irregularly filled desert depressions are remarkably well adapted for reproducing successfully whenever the opportunity is presented.

Artificial Waters as Routes of Range Expansion

Rzoska (1966) wrote that reservoirs of the USSR serve as "stepping stones" for the southward spread of northern species. The same applies on this continent. The northern cladoceran *Leptodora kindtii* has appeared in impoundments in southern Louisiana, Arkansas, Texas, Oklahoma, and Kansas (Conner and Bryan 1983). It seems to be moving southwestward. Similarly, the appearance of *Diaptomus reighardi* in Lake Mead (Paulson, Baker and Deacon 1980) represents a southwestward move for this planktonic copepod. Belk (1977a) discerned an opposite trend in small southwestern stock tanks. The two anostracans *Thamnocephalus mexicanus* and *Streptocephalus mackini* seem to be moving northwestward from Mexico via artificial waters, represos, and roadside ditches.

The red crayfish *Procambarus clarkii* now is common in canals and impounded waters far west of its original range. As a consequence of multiple introductions it has a "pantemperate-subtropical distribution" (Hobbs 1984). It is not the only crayfish species to be found in, for example, artificial waters of Arizona and California at great distances from ancestral homes, but it is most easily recognized by non-specialists.

Endemic Species and Artificial Waters

Because of the insular nature of waters in arid lands they sometimes contain endemic species. Impoundment, of course, modifies and reduces much of the natural habitat to the detriment of these forms. Some, however, can tolerate impoundment more than they can the purposeful introductions of exotic species that follow the creation of a reservoir. Just one example is the razorback sucker (*Xyrauchen texanus*), an endemic to the Colorado River basin. Now it is found only as scattered individuals except for a population of old, slowly growing fish in Lake Mohave of Arizona and Nevada. Much evidence supports the idea that predation on the embryos, larvae, and juveniles of the razorback by introduced fishes is the primary reason for its decline (Minckley 1983). Parenthetically, of 29 species of fishes collected through the years from Lake Mohave, only 6 were native forms. Such a ratio typifies many reservoirs in the arid West.

We can look forward to the discovery of undescribed species in reservoirs of desert and semi-arid regions. These will have been introduced in some manner from natural waters where they existed as endemic species or as forms with very limited distribution. The occurrence of an unusual cyclopoid copepod, morphologically referable to *Acanthocyclops vernalis* (Fisher), in two small, lined ponds in Phoenix, Arizona, may be typical. This form, probably a sibling species of *A. vernalis*, appeared in the artificial ponds and was studied by Foster (1973) as part of a broader ecological investigation. Five generations were produced during the summer months, each one marked by the first adults being 100% males, and the close of each generation being composed of 100% females. The adult males passed spermatophores to the copepodid-V females; no adult females ever were found with spermatophores as is typical. This unusual cyclopoid mating behavior may be a product of selection for survival in short-lived desert pools. The probability is great that Foster's copepods are reproductively isolated from other populations that would "key out" to *A. vernalis*. The situation probably is not unique; other unusual plants and animals can be anticipated in southwestern reservoirs.

Plankters, Planktivores and Food Webs

Power and water requirements of cities and ranches and the resultant "drawdowns" cause marked changes in reservoir water levels. At times the lakes are drained almost completely, thereby interrupting any smooth successional trends. This applies especially to the composition of zooplankton populations. Larger, opaque cladocerans and copepods are scarce or absent when fish are present, particularly planktivores such as the species of *Dorosoma*. Following the drainage of Saguaro Lake, Arizona, and the removal of fish, the large, pigmented *Daphnia pulex* became abundant in the limnetic zone; before this it was rare and restricted to shallow water near the bottom. It disappeared from the plankton assemblage when fish were introduced. Similarly, McNatt (1977) found that *D. pulex* persisted at the time Canyon Lake was being refilled in early 1970, but disappeared by 1971. He noted also that thriving threadfin shad populations resulted in a zooplankton population dominated by small rotifers, cyclopoid copepods, and the cladoceran *Bosmina*. Earlier, Cole (1961, 1963) reported the large calanoid copepod *Diaptomus clavipes* and the smaller *D. siciloides* from the Salt River reservoirs. McNatt found only the latter when threadfin populations were high. Paulson, Baker, and Deacon (1980) reported the small *D. siciloides* to be the most abundant calanoid in Lake Mead, although some *D. clavipes* were also present.

Because some reservoirs can be drained, it is possible to experiment further with the concept of vertebrate predators modifying the composition of zooplankton populations. Hrbacek (1962) and Brooks and Dodson (1965) were the first to show that large, pigmented forms are replaced by smaller, transparent species when planktivorous fish are introduced. Few workers in the field have followed closely the events occurring before and after a reservoir is drained, with the subsequent introduction of fish.

In steep-sided reservoirs of arid zones, the absence of an abundant littoral-sublittoral biomass is a factor in sports fishery. Rich plankton productivity is at the base of the trophic structure in these lakes and, without a pelagic planktivore, much of this is denied such popular game fishes as the largemouth bass (*Micropterus salmoides*). The threadfin shad (*D. petenense*) has been a link between plankton production and the game fishes both in Arizona canyon impoundments on the Salt and Verde rivers and in Lake Mead. The decline of threadfin populations in Lake Mead since the closing of Glen Canyon Dam, 450 km upstream, is probably associated with a lowering of water fertility. The ultimate result is a serious decrease in the number of largemouth bass and striped bass (*Morone saxatilis*) in the lake.

PUNCTUATED ASPECTS OF SOME ARID-LAND RESERVOIRS

Meterologic events occurring above impoundments and on their watersheds in deserts and prairies are irregular, localized, and difficult to predict. The long calm period punctuated by the violent storms that affect limited areas is typical. The flash flood is a dominant arid-zone phenomenon. The probability of precipitation and/or wind may be greatest in a particular season and vernal snowmelt can be predicted with some accuracy, but, compared to a more mesic region, uncertainty prevails in semi-arid and arid lands.

There is a variegated aspect to some arid-land reservoirs, especially those located in steep desert canyons. The lengthy stretch of river that is impounded to form such a lake is joined by diverse tributaries. There are at least 17 tributaries in the 450-km reach of the Colorado River between Lees Ferry and Lake Mead. Some are major streams with extensive drainage areas and gentle gradients; of these, a few are intermittent, varying seasonally, and at times they respond with enormous swollen floods carrying heavy loads of sediment. The most common type of tributary, however, is a short creek with a steep gradient, carrying clear water at low discharge; when flooded, however, such creeks can sweep huge boulders to the main stream.

Now that it appears the threat of a new impoundment in this reach of the Colorado River is past, these tributaries will not be flooded by the rising waters of a new lake to which they would contribute their flow. Whereas the average discharge of each of these streams is only about 700 cfs (1,191 m^3/min), their combined maximum flow is 72 times this amount. The lithology of their drainage areas and/or aquifers results in five distinct water types on the basis of ionic composition. These affect different sections of the river, raising salinity in some places and subsequently diluting it in others. These effects might not be obvious in a huge reservoir, but certainly the extreme turbidity introduced by the Little Colorado river in times of flood would be an obvious point source in a riverine lake.

Many geologic strata are based in deeply incised canyons. The vertical and longitudinal differences in the bed of a lake formed by impounding such a canyon cannot be matched elsewhere. Millions of years of sedimentary and erosional history are involved. Granites, schists, limestones, shales, and sandstones make up variegated basins that cannot be found in humid regions. Perhaps these factors and the complicated morphology of riverine impoundments account, in part, for spatial and temporal variations in their planktonic community compositions, standing crops, and productivities. Silvey and Stanford (1978) generalize that significant variations in pelagic communities are less pronounced in natural lakes than in reservoirs of the southwestern United States.

One of the most engaging of the punctuated, interrupted features of some desert stream lakes, e.g., Lake Powell, deals with nutrient sources. As each tributary is invaded, old alluvial terraces are flooded and nutrients are released. Here also a mesic riparian vegetation may develop and serve as a source of nutrients in contrast to the extensive bare cliffs along the lake on both sides of the inundated tributary. Near the mouths of some entering streams, lacustrine deposits from old beaver dams are present, contrasting as important nutrient sources with adjacent bare shores. Another zoogenic focal point of nutrient release is the crevice or protected ledge that supported packrats' or birds' nests. Organic matter has collected in these locations, protected from the rain, for centuries or perhaps millennia. These dry nitrogenous collections support luxuriant benthic algal masses after being flooded. Potter and Pattison (1976) pinpointed these rich sources in Lake Powell by tracing the terminal dendritic pattern of the algae upslope as it changed to an ever-greener linear path, 1–2 m wide and a centimeter thick, leading directly to the ancient nest.

REFERENCES

Anderson, E. R., and D. W. Pritchard. 1951. *Physical Limnology of Lake Mead.* U.S. Navy Electronics Laboratory Report 258. San Diego, California.

Barry, W. T. 1980. Personal communications with the author.

Bayly, I. A. E., and W. D. Williams. 1973. *Inland Waters and Their Ecology.* N.p.: Longman Australia Pty Limited.

Belk, D. 1977a. "Zoogeography of the Arizona Fairy Shrimps (Crustacea: Anostraca)." *Journal of the Arizona Academy of Science* 12(2):70–78.

———. 1977b. "Evolution of Egg Size Strategies in Fairy Shrimps." *The Southwestern Naturalist* 22(1):99–105.

Brooks, J. L., and S. I. Dodson. 1965. "Predation, Body Size, and Composition of Plankton." *Science* 150:28–35.

Cantwell, R. J. 1965. *An Analysis of the Limnology of Ganado Lake.* Master's thesis, Arizona State College, Flagstaff.

Cole, G. A. 1961. "Some Calanoid Copepods from Arizona with Notes on Congeneric Occurrences of *Diaptomus* Species." *Limnology and Oceanography* 6(4):432–42.

———. 1963. "The American Southwest and Middle America." In *Limnology in North America,* ed. D. G. Frey, 393–434. Madison: University of Wisconsin Press.

———. 1966. "*Branchiura sowerbyi* Beddard (Annelida: Oligochaeta) in Arizona." *Journal of the Arizona Academy of Science* 4(1):43.

———. 1983. *Textbook of Limnology.* 3d ed. Saint Louis, Mo.: The C. V. Mosby Company.

Conner, J. V., and C. F. Bryan. 1983. "*Leptodora kindtii* (Focke) (Crustacea: Cladocera) in Southern Louisiana Since 1973." *The Southwestern Naturalist* 28(1):118–19.

Deacon, J. E., and W. L. Haskell. 1963. "Occurrence of the Freshwater Jellyfish, *Craspedacusta sowerbyi,* in Lake Mead, Nevada." *The American Midland Naturalist* 70(2):504.

de Jong, R. L. A. 1973. "Fall and Rise of Lago del Oro." In *Man-made Lakes: Their Problems and Environmental Effects,* ed. W. C. Ackerman, G. F. White, and E. B. Worthington, 292–94. Geophysical Monograph, Vol. 17. N.p.

Dendy, J. S. 1945. "Fish Distribution, Norris Reservoir, Tennessee, II-Depth Distribution of Fish in Relation to Environmental Factors." *Journal of the Tennessee Academy of Science* 20(1):114–35.

———. 1963. "Farm Ponds." In *Limnology in North America,* ed. D. G. Frey, 595–620. Madison: University of Wisconsin Press.

Dodson, S. I., and S. D. Cooper. 1983. "Trophic Relationships of the Freshwater Jellyfish *Craspedacusta sowerbyi* Lankester 1880." *Limnology and Oceanography* 28(2):345–51.

Eakin, H. M., and C. B. Brown. 1939. *Silting of Reservoirs.* U.S. Depatment of Agriculture, Technical Bulletin Number 524. Washington, D.C.

Ellis, M. M. 1940. "Water Conditions Affecting Aquatic Life in Elephant Butte Reservoir." *Bulletin U.S. Bureau of Fisheries* 49:257–304.

Evans, K. J. 1959. "A Record of the Oligochaete *Branchiura sowerbyi* from Oklahoma." *The Southwestern Naturalist* 3(1958):223–24.

Fisher, H. B., and R. D. Smith. 1983. "Observations of Transport to Surface Waters from a Plunging Inflow to Lake Mead." *Limnology and Oceanography* 28(2):258–72.

Foster, J. M. 1973. *Limnology of Two Desert, Recharged-Groundwater Ponds.* Ph.D. diss., Arizona State University, Tempe.

Freeze, R. A., and J. A. Cherry. 1979. *Groundwater.* Englewood Cliffs, N.J.: Prentice Hall Inc.

Geraghty, J. G., and D. W. Miller. 1973. *Water Atlas of the U.S. Water Information Center.* Port Washington, N.Y.

Gerking, S. D. 1963. "Central States." In *Limnology in North America*, ed. D. G. Frey, 239–68. Madison: University of Wisconsin Press.

Grover, C., and C. S. Howard. 1938. "The Passage of Turbid Water Through Lake Mead." *Transactions of the American Society of Civil Engineers* 103:720–90.

Harris, B. B., and J. K. G. Silvey. 1940. "Limnological Investigation on Texas Reservoir Lakes." *Ecological Monographs* 10(1):111–43.

Hobbs, H. H., Jr. 1984. "On the Distribution of the Crayfish Genus *Procambarus* (Decapoda: Cambaridae)." *Journal of Crustacean Biology* 4(1):12–24.

Hoffman, D. A. and A. R. Jonez. 1973. "Lake Mead, a Case History." In *Man-made Lakes: Their Problems and Environmental Effects*, ed. W. C. Ackermann, G. F. White, and E. B. Worthington, 220–33. Geophysical Monograph, vol. 17. N.p.

Hrbacek, J. 1962. "Species Composition and the Amount of Zooplankton in Relation to the Fish Stock." *Rozpravy Ceskoslovenske Akademie Ved Rada, Mat. Prir. Ved* 72(1):1–116.

Hutchinson, G. E. 1937. "A Contribution to the Limnology of Arid Regions Primarily Founded on Observations Made in the Lahontan Basin." *Transactions of the Connecticut Academy of Arts and Sciences* 33(1):47–132.

———. 1951. "Copepodology for the Ornithologist." *Ecology* 32(3):571–77.

Hutchinson, G. E., G. E. Pickford, and J. F. M. Schuurman. 1932. "Contribution to the Hydrobiology of Pans and Other Inland Waters of South Africa." *Archiv fur Hydrobiologie* 24(1):1–154.

Jester, D. B. 1971. "Effects of Commercial Fishing, Species Introductions, and Drawdown Control on Fish Populations in Elephant Butte Reservoir, New Mexico." In *Reservoir Fisheries and Limnology*, ed. G. E. Hall, 265–85. American Fisheries Society, Special Publication Number 8. Washington, D.C.

Koenig, E. R., R. W. Tew, and J. E. Deacon. 1972. "Phytoplankton Succession and Lake Dynamics in Las Vegas Bay, Lake Mead, Nevada." *Journal of the Arizona Academy of Science* 7(3):109–12.

Kubly, D. M., and G. A. Cole. 1979. "The Chemistry of the Colorado River and Its Tributaries in Marble and Grand Canyons." In *Proceedings of the First Conference on Scientific Research in the National Parks*, ed. R. M. Linn, 565–72. U.S. Department of the Interior, National Park Service Transactions and Proceedings Series vol. 1, no. 5. N.p.

Kynard, B. E., and J. C. Tash. 1974. "Freshwater Jellyfish (*Craspedacusta sowerbyi*) in Lake Patagonia, Southern Arizona." *Journal of the Arizona Academy of Science* 9(2):76–77.

Lind, O. T. 1971. "The Organic Matter Budget of a Central Texas Reservoir." In *Reservoir Fisheries and Limnology*, ed. G. E. Hall, 193–202. American Fisheries Society, Special Publication Number 8. Washington, D.C.

Margalef, R. 1975. "Typology of Reservoirs." *Verhandlungen der Internationale Vereinigung fur Theoretische und Angewandte Limnologie* 19:1841–42.

McConnell, W. J. 1968. "Limnological Effects of Organic Extracts of Litter in a Southwestern Impoundment." *Limnology and Oceanography* 8(2):343–49.

McConnell, W. J., and G. F. Adams. 1973. "Emory Oak (*Quercus emoryi*) Litter Phenolics as Environmental Hazards for Aquatic Animals in Southeastern Arizona." *Journal of the Arizona Academy of Science* 8(3):111–15.

McConnell, W. J. and A. Longo. 1971. "Litter Phenolics of Possible Limnological Importance in Southeastern Arizona." *Journal of the Arizona Academy of Science* 6(4):235–37.

McNatt, R. M. 1977. *Zooplankton Species Composition, Abundance, and Distribution in a Desert, Pumped-Storage Reservoir*. Ph.D. diss., Arizona State University, Tempe.

Minckley, W. L. 1983. "Status of the Razorback Sucker, *Xyrauchen texanus* (Abbott), in the Lower Colorado River Basin." *The Southwestern Naturalist* 28(2):165–87.

Neel, J. K. 1963. "Impact of Reservoirs." In *Limnology in North America*, ed. D. G. Frey, 575–93. Madison: University of Wisconsin Press.

Neumann, J. 1953. "Energy Balance of and Evaporation from Sweet-Water Lakes of the Jordan Rift." *Bulletin of the Research Council of Israel* 2(4):337–57.

Paulson, L. J., J. R. Baker, and J. E. Deacon. 1980. *The Limnological Status of Lake Mead and Lake Mohave Under Present and Future Power-Plant Operations of Hoover Dam.* Lake Mead Limnological Research Center, University of Nevada. Nevada Technical Report Number 1. Las Vegas.

Pennak, R. W. 1963. "Rocky Mountain States." In *Limnology in North America*, ed. D. G. Frey, 349–69. Madison: University of Wisconsin Press.

———. 1968. "Historical Origins and Ramifications of Interstitial Investigations." *Transactions of the American Microscopical Society* 87(2):214–218.

———. 1971. "A Fresh-Water Archiannelid from the Colorado Rocky Mountains." *Transactions of the American Microscopical Society* 90(3):372–75.

Potter, L. D., and E. T. Louderbough. 1977. "Macroinvertebrates and Diatoms on Submerged Bottom Substrates, Lake Powell." Lake Powell Research Project Bulletin Number 37, University of California, Los Angeles.

Potter, L. D., and N. B. Pattison. 1976. *Shoreline Ecology of Lake Powell.* Lake Powell Research Project Bulletin No. 29, University of California, Los Angeles.

Rzoska, J. 1966. "The Biology of Reservoirs in the U.S.S.R." In *Man-made Lakes*, ed. R. H. Lowe-McConnell, 149–54. Symposia of the Institute of Biology Number 15. London: Academic Press.

Scholtz, J. F. 1968. *Geology of the Woods Canyon Drainage Basin, Coconino County, Arizona.* Master's thesis, Northern Arizona University, Flagstaff.

Silvey, J. K. G., and J. A. Stanford. 1978. "A Historical Overview of Reservoir Limnology in Southwestern United States." In *Current Perspectives on River-Reservoir Ecosystems*, ed. J. Cairns, E. F. Benfield and J. R. Webster, 1–18. Proceedings, Symposium of North American Benthological Society, April 1977, Roanoke, Va.

Ward, J. V. 1977. "First Records of Subterranean Amphipods from Colorado with Descriptions of Three New Species of *Stygobromus* (Crangonictidae)." *Transactions of the American Microscopical Society* 96(4):452–66.

Wiebe, A. H. 1938. "Limnological Observations on Norris Reservoir with Special Reference to Dissolved Oxygen and Temperature." *Transactions of the North American Wildlife Conference* 3:440–57.

3

Pond Limnology, with Applications to Arid-Area Ponds

William J. Clark

SUMMARY

Ponds, which can be defined as bodies of shallow water having small volumes, surface areas, and watersheds, are highly variable and therefore defy generalities and predictions. Nonetheless, one generally finds that their littoral zones are large, vascular plants are very influential, and stratification is variable. Oxygen deficits can occur, and there often are rapid changes in water temperature and chemistry. The individuality of ponds is enhanced by unevenly distributed winds and precipitation, diverse land use practices, and soil types. A study of five ponds in Brazos County, Texas, found that, like many of the variables measured, water temperature stratification was transitory. Although dissolved oxygen amounts varied and changes in concentrations were rapid and difficult to predict, all the ponds studied were undersaturated. In addition, there were no consistent patterns of Secchi-disc depth and turbidity, and there was no stability in alkalinity. Water chemistry profiles depended more on bottom material composition than on soil chemistry and were altered more by specific events than by daily changes. Among the ponds there was considerable interpond variability, but great intrapond consistency. Algal Growth Potential experiments showed that AGP looks promising for identifying regional limnological characteristics and variant ponds. Ponds in arid areas, most of which are man-made and include impoundments, those without watersheds, and excavated ponds, also exhibit great diversity. It is because of this individuality of ponds that the author believes carefully planned multi-pond studies could establish categories of ponds, but could never provide a high level of predictability for a given period.

INTRODUCTION

Understanding pond limnology is a difficult task because significant departures from classical lake limnology must be considered. Ponds in arid areas have the added complication of being sysems that are influenced by factors which are not major controlling factors in ponds in general.

This chapter presents some generalizations about ponds and the departures from classical lake limnology which must be made in studying them. Examples from Texas ponds and special characteristics of arid-area ponds are discussed.

WHAT IS A POND?

Professional limnologists have not been able to agree on any simple technical criteria which sets a water body apart as a *pond.* One group has even chosen to avoid the subject (Ruttner 1953, Macan 1974, Moss 1980, Cole 1983).

Forel (1892) defined a pond as "a lake of slight depth." Carpenter (1928) defined a pond as "a quiet body of water in which the littoral zone of floating leaved vegetation extends to the middle of the basin, and in which the biota is very similar to that of the littoral zone of lakes." Thus, the concept of shallow depth and the associated implications for plant growth were introduced at an early date.

Welch (1952, 15) discussed the problem in some detail. He reviewed the older literature and brought in such concepts as the presence or absence of a wind-swept shore and thermal stratification as definitive criteria. He not only pointed out the great diversity involved, but also the improbability of any single criterion being satisfactory. He concludes: "For limnological purposes it is possible that Forel's definitions are about as satisfactory as any now proposed. None seems to be wholly ideal."

Goldman and Horne (1983, 36) point out that if depth is accepted as the arbiter, Lake Chad and Lake Winnipeg should be classified as ponds, and that if vegetation is the criterion, all of the small, shallow arctic coastal waters would be lakes regardless of their small size. They also introduce a new and interesting concept. "Water bodies may be considered lakes when the wind plays the dominant role in mixing. In ponds the gentler convective mixing predominates." Convective mixing cannot form a discrete epilimnion, since it only operates when the pond is cooling. Wind energy is required to form an epilimnion in the spring. Their definition, therefore, would seem to restrict ponds to bodies of water which do not form thermal stratification.

In the only two books specifically about ponds (Boyd 1979; Dykyjova and Kvet 1978) the point does not come up, since both are concerned with man-made ponds used for fish culture. Both, however, contain considerable information about the function of pond ecosystems. What constitutes a pond in a particular area and which of the criteria is of particular importance in the ecology of that pond must be left up to the judgment of a limnologist familiar with the area. One fact is clear: the number of small bodies of water (natural and man-made) is enormous, warranting special attention.

In an inventory study now being prepared for publication, estimates were made that there are about 190,000 ponds smaller than 1.25 ha in the Brazos River basin in Texas, and about 102,000 in the Colorado River basin. The total number for Texas, based on densities arrived at in these two basins, is 782,000 ponds smaller than 1.25 ha, and 844,000 smaller than 11.25 ha.

GENERAL CHARACTERISTICS OF PONDS

The following characteristics apply to most but not all ponds. All of the characteristics, however, have important limnological implications which must be kept in mind if ponds are to be understood. Categorically, ponds are relatively shallow and have small volumes, surface areas, and watersheds.

Shallow Depth

The littoral region (the shore region in the photic zone where rooted aquatic plants can grow) is large in relation to the total area of the pond and in clear ponds may cover the entire bottom of the pond. Vascular plants and the organisms associated with them thus play a much more important role in the total productivity of ponds than is true in most lakes. Where significant beds of vascular plants are present, total pond productivity can be estimated only by using whole pond methods. Standard light bottle–dark bottle methods may be grossly misleading.

Stratification in shallow ponds is highly variable and is dependent upon the interaction of depth, climate, and shoreline factors. Several classes of stratification can be recognized:

1. Rare stratification with water temperatures reacting quickly to wind action and convective mixing (very shallow).

2. Irregular stratification often stratifying during calm periods, but mixing easily when wind increases.

3. Incomplete stratification with a fairly discrete epilimnion and metalimnion, but not deep enough to develop a true hypolimnion. The term *metalimnion* is used here for the transition zone between the epilimnion and the hypolimnion, in preference to the term *thermocline* for which there is some confusion as to the exact definition and proper usage.

4. The classical stratification pattern (hypolimnion nearly isothermous with depth and changing little during the stratification period) develops in deeper ponds.

Oxygen and nutrient stratification are closely linked to density stratification in ponds of Class 4. Although infrequently realized, oxygen and nutrients can be stratified in any of the density stratification classes discussed above. Low oxygen values in summer may be found in ponds which have no strong thermal stratification, only an erratic decrease in temperature with depth, or in ponds which have the epilimnion-metalimnion stratification. These latter ponds may exhibit a gradual warming of the metalimnion with time, evidencing the lack of isolation of the deeper water. The basic requirement for an oxygen deficit, for example, is simply an imbalance between the rate of respiration and the rate of downward transport of oxygen and of oxygen supply by photosynthesis. Since the water volumes are relatively small, the amount of organic material necessary to cause the oxygen deficit is not large, and the deficit can develop in a very short time.

Small Volumes

As discussed above, small volumes contribute to oxygen deficit problems. Small volumes also mean small heat-holding capacity, and for this reason changes in water temperature can be rapid. Because storm runoff volumes may be greater than the pond volume in some cases, complete flushing of the pond can occur, which in turn causes rapid fluctuations in water chemistry and losses of pelagic populations of biota.

Small Surface Area

The wind speed and the length of the open water over which the wind can blow (the fetch) determine the amount of energy that can be transmitted to the water. For a given wind speed, as the fetch becomes shorter, the waves become smaller and mixing decreases. The presence of beds of submerged aquatic plants reduces the mixing, and emergent vasculars may provide some windbreak, but brush or trees on the bank will be more efficient. If the pond is constructed in a gully or watercourse, or if the excavated material is piled along the shore, significant attenuation of the wind velocity can occur. Differences in the extent, nature, and orientation of these factors in relation to the prevailing winds and the orientation of the long axis of the pond can cause significant pond-to-pond individuality.

Small Watersheds

Winds and precipitation are very unevenly distributed, even in general storms, and thunderstorms have very sharply defined tracks. This means that ponds in the same general area may be very differently affected by the same storm. A thunderstorm may leave a swatch of ponds with added water and considerable mixing while ponds on both sides of the track remain unaffected. Small watersheds encompass fewer soil types and land use practices than large watersheds. Land use may change at the property line in ways that could significantly affect pond conditions in adjacent ponds.

Great care must be taken in interpreting data from pond studies so that differences caused by the factors discussed above are not mistaken for treatment effects.

BRAZOS COUNTY PONDS

The effects of some of the factors discussed previously will be illustrated with data from ponds in Texas, largely from Brazos County near Texas A & M University where synoptic studies on groups of ponds have been done for several years. These data most likely are not representative of ponds in general. Ponds in other areas of the country will have their own characteristic reactions to the basic parameters. The extent to which these data can be extrapolated even to other areas in Texas is still uncertain. This is the subject of current research. The data illustrate some of the basic points and emphasize the variation that occurs.

Temperature

Meyers (1973) studied temperature characteristics of several ponds including five which were within a quarter of a mile of each other, under similar land use practices, and on the same slope. Physical parameters for the ponds are given in Table 3.1. Water temperature profiles (Fig. 3.1a, b, c, d) illustrate the changes that occur during different seasons.

Table 3.1. Physical parameters of five experimental ponds in Brazos County, Texas

Parameter	Pond Number 1	2	3	4	5
Maximum depth (m)	1.8	2.3	2.1	9.1	4.6
Mean depth (m)	1.09	0.92	1.01	3.72	1.94
Fetch (m)	30	96	60	239	263
Volume (m^3)	495	1,630	945	74,604	14,266
Area (m^2)	455	1,770	940	20,079	7,352

Summer. Pond 5 was the clearest, and had the longest fetch. It was warmer, and formed the deepest epilimnion. The other four ponds had similar temperatures down to a depth of 1.5 m, though they maintained no fixed order. Figure 3.1a shows fairly well developed epilimnia, and Figure 3.2 taken six days later shows how transitory the stratification can be, even in summer.

Assuming these data are representative and extrapolatable, small ponds under 3 m deep in central Texas will have epilimnion depths of 0.5 to 1 m when they stratify, and larger ponds from 5 to 8 m deep will have epilimnia of 1.4 to 2 m. In comparison, flood retention reservoirs of near 12 ha area have epilimnia 2–2.4 m deep, and larger reservoirs such as Lake Conroe and Toledo Bend Reservoir have epilimnia 5–7 m deep.

Fall. Fall cooling sequences which emphasize pond individuality are shown in Figure 3.1b. Pond 5, which was the warmest, cools the slowest. The other ponds do not appear to segregate in any simple relationship to the parameters in Table 3.1.

Winter. Figure 3.1c shows a representative winter profile. The ponds do not appear to segregate in any more predictable manner when cooling than they did when warming. Pond 5, though the warmest in summer, was not the coolest in winter. The only pond to cool to the 4° maximum density point was Pond 2 which was intermediate in depth, fetch, and volume. Influences of shelter and orientation appear to have played a strong role. It has been said that central Texas does not have winter: it has "fronter." When the storm fronts go by, the weather can shift dramatically and quickly. Marked changes in temperature profiles can result from these abrupt climatic changes.

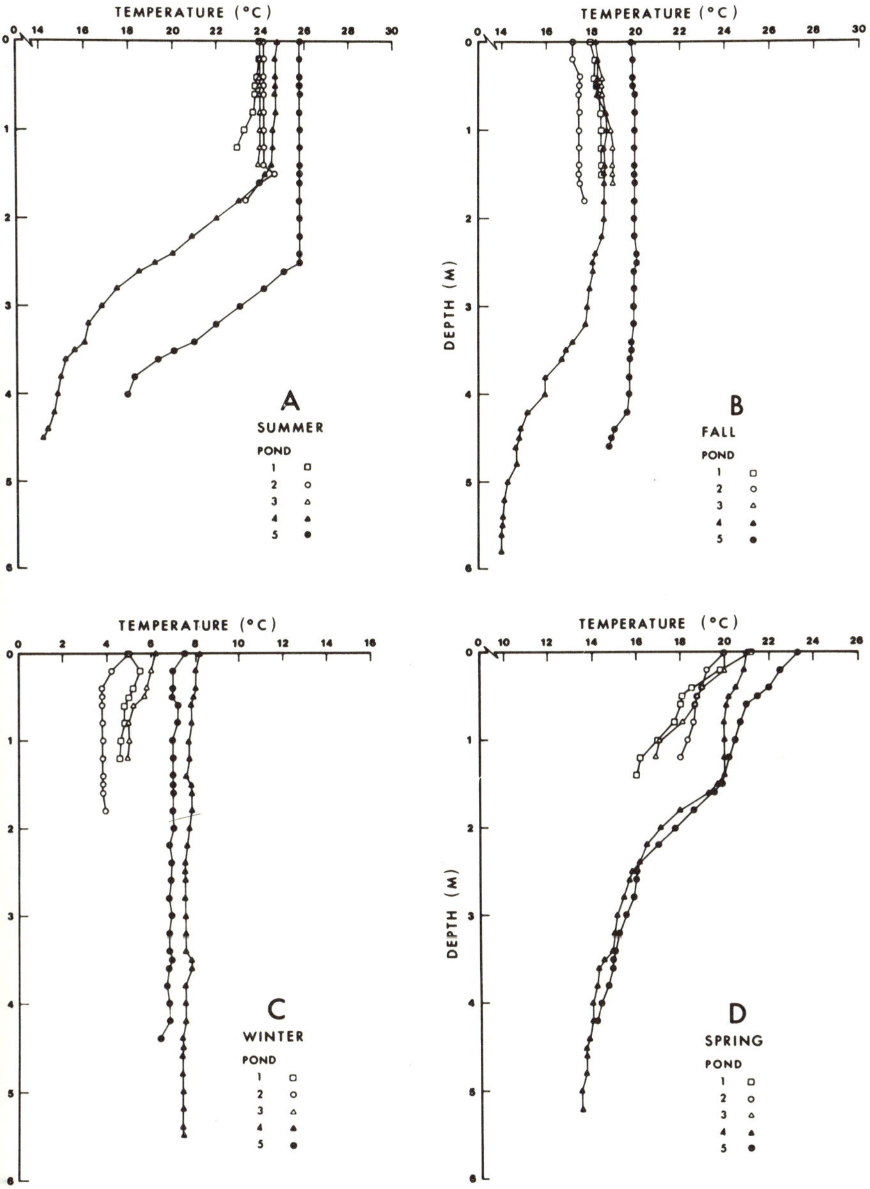

Figure 3.1. Representative temperature profiles for five experimental ponds in Brazos County, Texas, during summer, fall, winter, and spring.

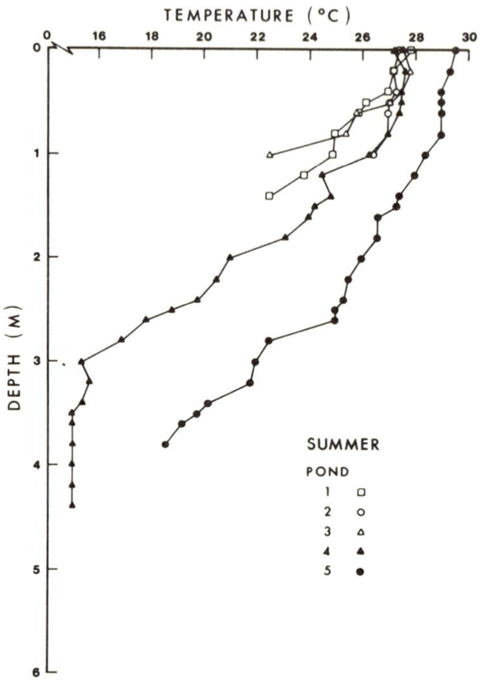

Figure 3.2. Temperature profiles six days after data collection for Figure 3.1a showing rapid change in stratification during summer.

Spring. Figure 3.1d shows a typical spring warming sequence. The ponds react individually, with Pond 5 already showing its superior heat-retaining capability.

A limnologist would have a difficult task trying to select any one of these ponds as "typical" of the area. Had the study been done on any one of them the generalizations drawn could have been misleading.

Oxygen

Unlike temperature, where the critical exchanges take place only at the air-water interface, internal biological activity at all depths moderates oxygen distribution. Dissolved oxygen is so intimately connected to biological processes that pond-to-pond variability is expected.

Summer. Dissolved oxygen (D.O.) varied widely from pond to pond. This variance was shown to be as great as 3.5 to 4.0 mg/l at a specific depth (Fig. 3.3). Most likely, the biological activity within each pond was sufficiently different to account for the D.O. variance. All of the ponds were undersaturated. Epilimnetic waters contained the highest concentrations of oxygen ranging from 2.5 to 6.2 mg/l. The deeper ponds, 4 and 5, usually exhibited the highest values. In Ponds 4 and 5 where a hypolimnion developed, D.O. was very low (0.4 to 1.5 mg/l). Abrupt intermittent climatic changes can markedly alter these patterns within a time span of a few days.

Fall. Figures 3.4a and b show a shift from slight D.O. supersaturation on October 1 to considerable undersaturation on October 29, coincident with a 6–8° C drop in surface temperature. These data illustrate the rapid changes that can result from temperature declines during this season.

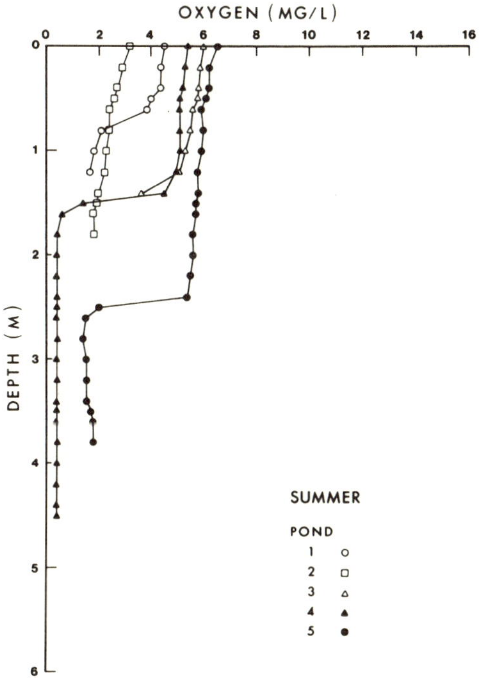

Figure 3.3. Representative oxygen profile for five experimental ponds in Brazos County, Texas, during summer.

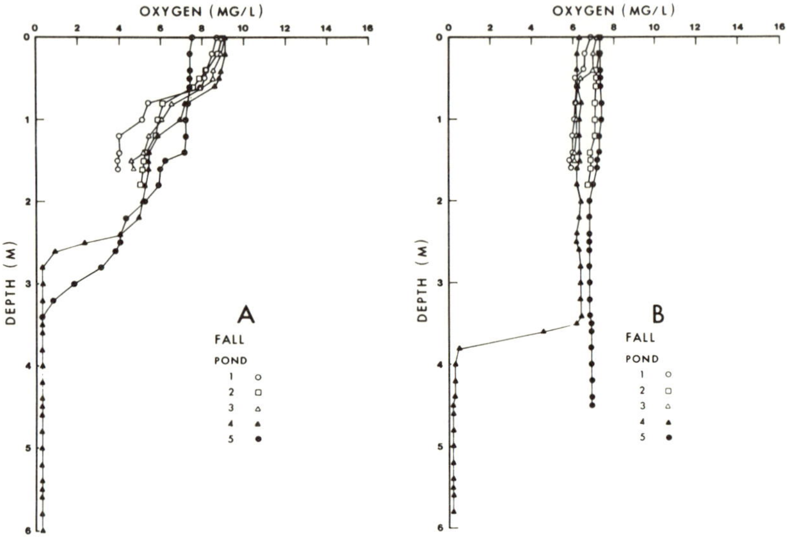

Figure 3.4. Variation in oxygen profiles for five experimental ponds in Brazos County, Texas, during fall: **a,** data collected on October 1; **b,** data collected on October 29. Comparison of these data illustrates the rapid changes that can result from temperature declines during the fall.

Winter. Colder water temperatures and reduced biological activity within the ponds result in increased dissolved oxygen concentrations. Maximum values were slightly over 11.0 mg/l in the epilimnion. In the shallow ponds—1, 2, and 3—the lowest D.O. was 8.4 mg/l. In the deeper waters of Pond 5, the D.O. dropped below 7.0 mg/l (Fig. 3.5).

The most significant point of these seasonal D.O. data is that in shallow, small volume systems when biological variability is imposed on physical parameter variability, changes in oxygen concentration can be rapid and difficult to predict.

Chemistry

Seasonal measurements of Secchi-disc depth and turbidity for the five ponds show high variability and no consistent pattern (Fig. 3.6). Ponds 3, 4, and 5 had watershed stability and show consistent turbidity values. Road construction in the watersheds of Ponds 2 and 3 had a significant influence

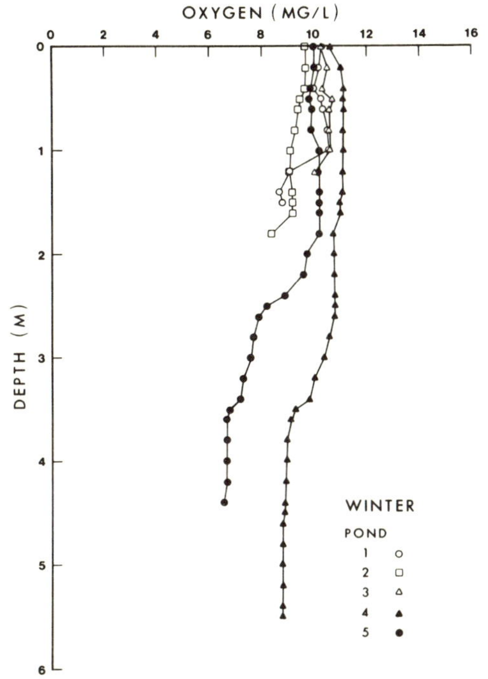

Figure 3.5. Representative oxygen profiles for five experimental ponds in Brazos County, Texas, during winter.

on the ponds, even though the activity occurred on a very small portion of the watershed. Recovery within the year is apparent in each case.

Total alkalinity, which might be expected to be a conservative measurement, shows no such stability. Ponds 1 and 5 varied the most and displayed different patterns (Table 3.2). Although alkalinity levels were in ranges that provide low to moderate buffering, pH levels changed little (Table 3.2). Year-to-year trends were similar in all of the ponds. The effects of biological processes on pH within these ponds were minimal.

The chemistry of pond water would be expected to have some relationship to the chemistry of the soil over which the runoff travels. Groundwater inflow will complicate this relationship to some extent, depending upon the relative volumes from the two sources. Statewide, it would be of great assistance if generalizations about pond water chemistry could be made from soil group characteristics; since regional soil characteristics are

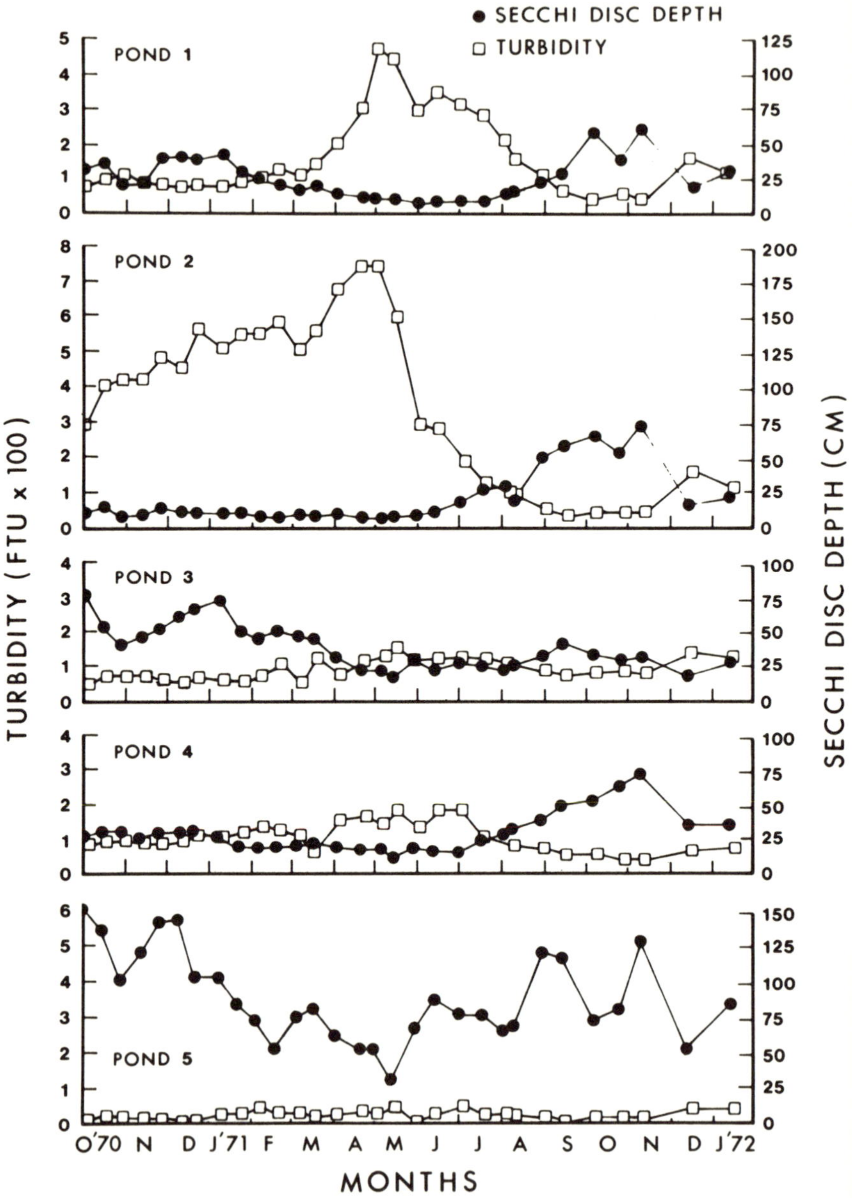

Figure 3.6. Season fluctuations in Secchi-disc depth and turbidity in five experimental ponds in Brazos County, Texas.

Table 3.2. Alkalinity and pH of the ponds studied

		Total Alkalinity (mg/l)			pH		
		1970	1971	1972	1970	1971	1972
Pond 1	Mean	17.5	17.6	17.93	7.3	6.6	6.89
	Range	14–24	10–32	6–32	6.8–8.2	6.3–7.8	6.3–7.4
Pond 2	Mean	21.5	36.95	33.0	7.52	7.08	7.16
	Range	18–30	16–54	10–46	7.0–8.1	6.7–7.7	6.7–7.6
Pond 3	Mean	24.0	24.95	29.19	7.40	7.3	6.97
	Range	20–34	14–38	18–62	6.7–8.3	6.6–8.6	6.1–7.8
Pond 4	Mean	22.75	27.00	38.22	7.54	7.33	7.13
	Range	20–36	20–40	20–64	6.9–8.6	6.8–8.1	6.5–8.1
Pond 5	Mean	27.7	30.43	51.04	7.59	7.46	7.23
	Range	24–32	22–50	24–136	7.1–8.5	7.1–8.2	6.7–8.4

different, such a generalization does not appear feasible. An examination of a series of ponds on clay soils in several areas where groundwater was not a factor was made by Springer (1983). From 40 to 60 soil samples were taken from each watershed, pooled, and their chemistry studied, along with that of the pond water and of the pond bottom sediment. Neither absolute chemical characteristics of the soils nor such factors as exchange capacity were well correlated with pond water chemistry. The closest relationships were between the pond bottom material and the pond water chemistry. One might hypothesize that the major part of the exchange in these systems takes place with the particles picked up in the stream channels, not with the undisturbed watershed soils. This would cause ponds in the same soil type to vary in their chemistry according to the extent and nature of the channel erosion.

One generalization about the way the chemical factors react in ponds is that except for periods of convective cooling, vertical profiles are altered by specific events far more than they are by day-to-day changes. Periods of daily sampling have shown that for several days there will be little change, and then a windstorm, thunderstorm, or frontal passage will cause drastic changes. This will be followed by another period of stability.

Biology

Biological diversity is a phenomenon which needs no validation. Laboratory microcosms filled with the same spring water and innoculated from the same source develop biological populations which diverge spectacularly. In aquaculture facilities, ponds are constructed at the same time, in the same oil, with the same equipment, and filled at the same time, from the same water source. Even so, they diverge as spectacularly in their biota as do the microcosms. Most quantitative biological studies are done on a single water body; however, in the light of the generalizations about biological diversity discussed above, how well can such quantitative single-pond studies represent the general area from which they are collected? This question remains unanswered for ponds generally, but it can be demonstrated that for the littoral of ponds in central Texas such a generalization could be grossly misleading.

A quantitative study of periphytic microcrustaceans (Campbell 1983) showed considerable interpond variability, but a high order of intrapond consistency. One representative example is given in Figure 3.7. Samples were collected from near the surfaces of *Ludwigia* plants, using an aspirator system. Data are reported for one of the more common forms, *Chydorus brevilabris*. Two of the ponds had two weed beds each, and the populations in the two weed beds in any one pond reacted almost identically to

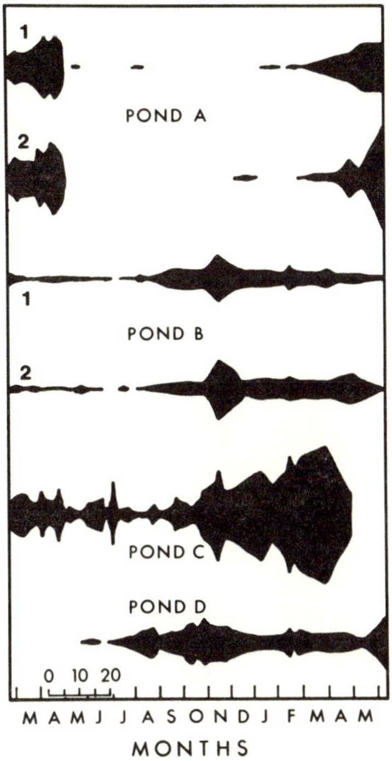

Figure 3.7. Seasonal variation in number of *Chydorus brevilabris* in four Brazos Bounty, Texas, ponds. The vertical width of the envelope, in scale units, represents the square root of the number of organisms per square decimeter of surface on *Ludwigia* plants. Ponds A and B each had two plant beds, while Ponds C and D had only one.

the parameters controlling population development. Pond-to-pond differences in seasonal patterns were spectacularly different, however, and none of the ponds would have given an accurate regional picture. Although all of the ponds were not as close together as the five used in the physical-chemical discussion, the two ponds with multiple weed beds were very close together.

Comparisons of aquatic systems are commonly made by studying integrating processes, such as primary productivity, or one or more of the critical components of productivity, such as chlorophyll or total phosphorus concentrations. If it is possible to study these two systems over time, such comparisons are effective. The study of only the planktonic primary productivity could be misleading in ponds with significant weed beds, and even in ponds without significant weed beds the physical-chemical parameters which affect productivity vary widely from day to day, and any single visit has a small probability of giving an appropriate long-term value.

Experiments are currently being made using the Algal Growth Potential (AGP) system, a modification of the EPA bottle test (Environmental Protection Agency 1971), in an attempt to find a procedure which would give a reliable long-term index of productivity on a single visit. The unicellular green algae *Selenastrum capricornutum* is being grown in triplicate in autoclaved water samples. Three replicates are taken for each sample using only an open light box on the laboratory table under ambient room temperatures. After the growth curve levels out, one sample is spiked with nitrate, one with phosphate, and one with a trace element mix. The study is still under way, and data on seasonal changes in the AGP are not yet complete. A representative set of data is presented in Figure 3.8 which illustrates both the promise and the frustration of the approach. Each line is the average of the three replicates for that pond up to the point of the spiking, after which only replicates which show some effect of the spiking are drawn.

Three of the four east Texas ponds reacted alike, but very differently from the Brazos County ponds. The fourth reacted very much like the Brazos ponds in maximum growth, but like the other east Texas ponds, it reacted only to the phosphorus spike. The rest of the east Texas ponds reacted only slightly or not at all to the phosphorus and did not react to the nitrate. Neither group reacted to the trace elements. Three of the Brazos County ponds reacted to nitrate but not phosphate, and the fourth reacted to a combined nitrate-phosphate spike given the next day. The procedure indicates general area differences, but illustrates the potential for pond individuality. AGP does look promising both for identifying regional limnological characteristics and for identifying variant ponds where detailed study might reveal the reasons behind their variant behavior.

ARID-AREA PONDS

The climatic factors which cause an area to be "arid" also dictate that there will be few permanent, natural bodies of standing water.

Very few persistent natural ponds exist in most arid areas, and they are a particularly important treasure, because they are often the last refuges of relic populations from past pluvial stages.

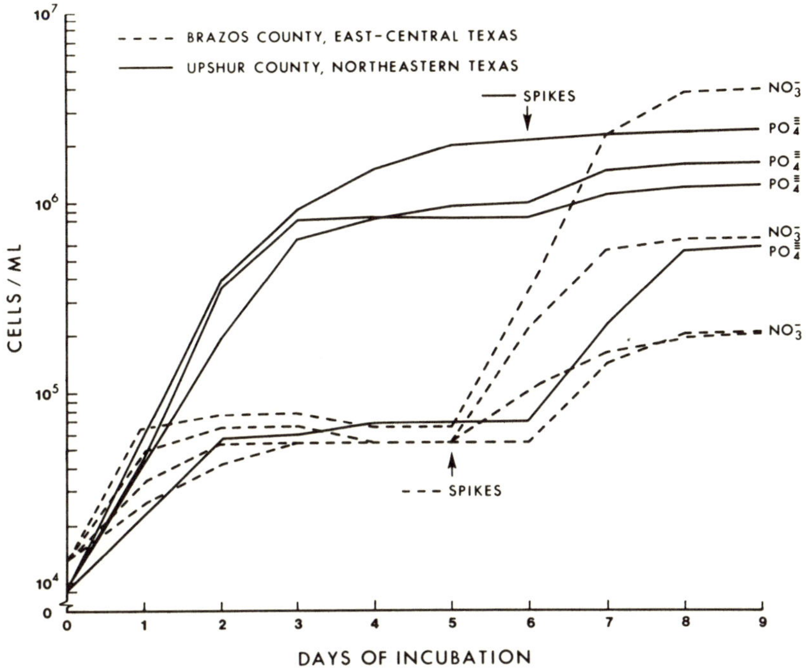

Figure 3.8. Growth curves for *Selenastrum capricornutum* innoculated into autoclaved pond water from ponds in Brazos and Upshur counties, Texas. Each line represents the average of three replicates up to the point of spiking. After growth leveled out, one replicate was spiked with nitrate, one with phosphate, and one with a trace element mixture. If spiking produced significant results, the lines from that point represent individual replicates and are labeled to show what the added nutrient was.

Most ponds in the arid areas are man-made, and these are the main object of this discussion. In general these ponds can be classified in three categories:

1. Impoundments, which intercept surface runoff.
2. Ponds without watersheds, which hold water from springs or pumped from wells.

3. Excavated ponds, which are cut below the water table but may also receive some runoff.

Impoundments are uncommon in truly arid areas. In some marginally arid areas, such as south Texas, there is still enough rainfall to permit ponds to persist during most years. These ponds rarely get enough inflow to cause them to overflow, and the evaporation rate is high. This results in large water level fluctuations and concentration of dissolved salts. They are primarily stock-watering ponds as well, and the effect of the stock usage on turbidity and nutrient supply is considerable. This group may be the best candidate for development of some area-wide characteristics.

Most arid-area ponds are ponds without watersheds, filled and maintained by water pumped from wells. Groundwaters are notoriously variable in availability and quality, and wells drilled near each other may vary even if drilled to the same depth. Volume and periodicity of flow are controllable, though this is practically never done on the basis of what is best for the biota. The most permanent ponds of this type are those used for livestock watering (primarily cattle), where a continuous water supply is an important requirement. These ponds provide an excellent laboratory to study the problems of dissemination of aquatic organisms, but the variability in chemical composition complicates the picture.

Excavated ponds are uncommon in arid areas because they require groundwater near the surface. In many arid areas the streams are aggrading in channels cut by much larger flows in the past. This provides the substrate for groundwater movement, and arid-area streams are noted for disappearing and then reappearing as surface streams, with the flow continuing as groundwater in the alluvium of the river valley. This is exemplified in the Pecos River valley in Texas. In several areas gravel operations have excavated large basins, which fill with groundwater. In this case the water is naturally saline, because of the saline formations in the upper watershed. The salinity approaches one-third to one-fourth of that of sea water, but with the ions in different proportions. The great variability in chemical parameters of arid-area waters is exemplified in Tables 3.3 and 3.4. Red drum have been grown successfully in them, and oysters also have survived and grown. They have a biota containing representatives of many of the common freshwater forms, which have adapted to these high salinities, but the systematics of the biota have not been well studied as yet.

The factors which dictate the nature of arid-area ponds add to the natural variability characteristic of ponds in general. The present state of our overall knowledge about ponds makes more detailed generalizations about arid-area ponds beyond our present capabilities. Pond research must make some progress first.

Table 3.3. Water analyses from west Texas wells and ponds

Pecos River Drainage	Nitrate (mg/l)	Chloride (mg/l)	Sulfate (mg/l)	Hardness (mg/l)	Bicarbonate Alkalinity (mg/l)	pH	Specific conductance (μ mho)
Wells, Pecos Co.							
1	1.8	43	50	260	182	7.2	590
2	0.8	290	525	780	200	7.3	2,900
3	12.5	2,500	2,375	2,890	210	7.2	8,300
4	1.5	3,500	3,500	3,650	53	8.8	9,800
5	4.5	575	1,375	1,360	182	—	—
6	12.0	1,780	2,250	2,520	197	7.1	6,200
7	14.0	150	—	560	138	—	—
8	1.2	95	—	290	164	—	—
Pecos River (FM1776)	1.3	6,750	3,250	3,350	122	—	—
Red Bluff Canal	0.7	3,500	3,000	2,360	69	8.1	9,500
Ponds, Pecos Co.							
Ford Pond	1.2	8,000	4,250	3,430	72	7.5	24,000
Phillips Gravel Pit	1.2	3,675	4,250	3,410	47	8.7	15,000
Porter Gravel Pit	1.3	7,450	4,250	3,320	110	8.1	17,500
Porter-Finley Gravel Pit	1.3	5,695	4,500	3,590	48	9.0	18,000
Rio Grande Drainage							
Neely Pond Hudspeth Co.	0.5	15,000	3,500	—	97	8.2	37,500

Table 3.4. Water analyses from west Texas, done by the Soil Testing Laboratory, Texas A & M University

	County Lab No.									
Analyses (mg/l except for pH and specific conductance)	Terry W8901	Ward W6955	Ward W4943	Ward W4945	Ward W4944	Ward W6946	Ward W6925	Ward W6927	Carnes W4972	Hudpeth W7693
Ca	424	518	995	1,015	767	912	864	874	77	853
Mg	2,125	287	444	710	369	505	547	541	9	3,395
K	240	5.5	61	49	40	26	25	7	29	920
Na	8,027	1,523	3,742	4,604	3,238	3,478	3,376	2,744	115	11,500
SO_4	13,442	2,231	2,954	4,046	2,711	8,353	3,375	4,050	433	11,439
Cl	7,300	2,104	6,417	7,701	5,501	5,538	4,596	3,877	144	13,591
Bicarbonate	812	274	118	95	142	121	83	150	—	143
pH	7.8	7.0	7.7	7.5	7.4	7.3	6.3	7.5	3.4?	6.8
Total Salts	32,371	5,700	13,198	15,673	12,539	10,488	10,488	8,956	8,836	25,077
Specific Conductance (μ mho)	33,707	9,193	21,288	25,280	20,224	16,853	16,853	14,445	1,348	40,448

CONCLUSION

Limnologists seem to be approaching the limit of effective contribution of single-pond studies to the solution of the important pond limnology questions. The great number of ponds, though part of the problem, can also be part of the solution. The opportunity for "replication" this provides is an advantage which can never be realized in most lake or reservoir studies. It is probable that carefully planned multi-pond studies will establish categories of ponds based on their various characteristics. An understanding of these categories will enhance our understanding of regional limnology of ponds, but may never provide a high level of predictability for any given period. Arid-area ponds suffer from the current, generally primitive level of our understanding of ponds and have the added problem of existing at the edge of the distribution. At this stage of the game they should be treated as individuals, and extrapolation to area characteristics should be made with great caution.

REFERENCES

Boyd, C. E. 1979. *Water Quality in Warm Water Rish Ponds.* Auburn, Ala.: Auburn University, Agricultural Experiment Station.

Campbell, J. M. 1983. *Interpond and Intrapond Variation in Populations of Periphytic Cladoceran Microcrustacea.* Ph.D. diss., Texas A & M University, College Station.

Carpenter, K. E. 1928. *Life in Inland Waters.* New York: N.p.

Cole, G. A. 1983. *Textbook of Limnology.* 3d ed. St. Louis, Mo.: C. V. Mosby.

Dykyjova, D., and J. Kvet. 1978. *Pond Littoral Ecosystems.* Berlin: Springer-Verlag.

Environmental Protection Agency. 1971. *Algal Assay Procedure: Bottle Test.* National Eutrophication Research Program, Corvallis, Ore.

Forel, F. A. 1892. *Monographie Limnologique.* Lausanne: N.p.

Goldman, C. R., and A. J. Horne. 1983. *Limnology.* New York: McGraw-Hill.

Macan, T. T. 1974. *Freshwater Ecology.* 2d ed. New York: John Wiley.

Meyers, D. G. 1973. *Comparative Physical Limnology of Farm Ponds in Southwest Texas.* Master's thesis, Texas A & M University, College Station.

Moss, B. 1980. *Ecology of Fresh Waters.* John Wiley, New York.

Ruttner, F. 1953. *Fundamentals of Limnology.* 2d ed. Toronto: University of Toronto Press.

Springer, T. A. 1983. *Relationships Between the Composition of Waters of East and Central Texas Lakes, and the Ion Exchange Characteristics of their Watersheds and Muds.* Master's thesis, Texas A & M University, College Station.

Welch, P. S. 1952. *Limnology.* 2d ed. NEW York: McGraw-Hill.

4

Fish Communities of Small Impoundments of Semi-Arid Areas

Richard L. Noble

SUMMARY

An important function of small impoundments in a multiple-use setting is providing fish habitats. Some factors affecting fish communities in ponds in semi-arid regions are the species composition and environmental influences such as water level, clarity, and fertility. The total number of species in small impoundments is usually few when compared to larger reservoirs. Stocking fish in combinations of predatory and prey species will attain a balance to ensure continued reproduction, to enhance size diversity, and to increase growth rates of both species. In semi-arid areas, the water level of the pond has the greatest effect on fish populations and can be used as a management tool. Water clarity must be adequate for photosynthesis and predation to occur, yet limited to prevent excessive macrophytic production. Introduced fertilization has little application in semi-arid ponds as nutrients are available from other sources. Ponds for fisheries can be managed successfully using the philosophy that a simple system is the most manageable, although pond management can include constructing single ponds with gently sloping shorelines and borrow pits and reclaiming and stocking ponds where feasible. Although small impoundments in semi-arid areas may not be conducive to habitat management, turbidity can be minimized by pond and bank stabilization; vegetation can be controlled chemically and biologically; and harvesting can be regulated through annual quotas and special size limits.

INTRODUCTION

Like large reservoirs, small impoundments are usually multiple-use resources. The most common of these, watershed ponds, are typically constructed to retain runoff for livestock drinking water. Such ponds are likely to be less than 2 ha in surface area. In those areas where erratic rainfall leads to downstream flooding, larger impoundments may be constructed as part of a comprehensive watershed project, e.g., floodwater-retarding structures, which average about 10 ha (3–20 ha). Besides these primary purposes, small impoundments serve other functions such as soil and water conservation and recreation. In recent years, construction of small impoundments principally for aesthetics and recreation, including fishing, has increased. Although the extent of pond construction and primary purposes of ponds vary regionally, fishing is usually an important use.

Because fishes are consumers (sometimes termed secondary producers), the biomass of fishes is dependent upon the level of limnetic primary productivity (phytoplankton) and the availability of lower level consumers (zooplankton, benthos, littoral microcrustacea, and even small fishes). Since fishes are long-lived, compared to most other biotic components of the aquatic ecosystem, their biomass is less subject to short-term variability, such as that caused by the change of seasons, than is the biomass of other organisms. Consequently they serve as integrators of the various components of the food web, and therefore the standing crop of fishes can be a reasonable indicator of pond productivity. With knowledge of the biology of principal pond fish species, the experienced scientist can frequently evaluate the trophic state of a pond more easily from a single determination of fish biomass than from repeated measures of chlorophyll levels, phytoplankton composition, or zooplankton densities.

In ponds, standing crop (biomass) at any given point during the year appears to be a reasonable indication of carrying capacity, at least for most adult fishes. Biomass of young fishes and frequently adults of small species will vary somewhat more, but the contribution of such cases to the total biomass of fishes will typically be negligible. Although standing crop likely varies directly with prey availability, standing crop indicates little about the dynamics of a pond system. Production, that is, biomass elaboration (Chapman 1967), is a measure of energy flow from the food web into the fish community which is then converted into new fish biomass. Production, calculated as the product of growth rate and mean biomass (though seldom estimated for pond populations), may range from nearly zero (no growth occurring) to values in excess of standing crop over a single growing season (Anderson 1973). Although estimates of fish production would be more meaningful in evaluation of trophic dynamics of ponds, standing crop (or an index of standing crop) is usually considered adequate for assessments of pond fish status.

The objective of this chapter is to discuss the principal factors affecting fish communities in small impoundments of semi-arid regions, with comparisons and contrasts to ponds of more humid areas. Relationships described in this paper are largely those developed or modified during studies of small impoundments in Texas. These studies spanned most areas of Texas with adequate rainfall to maintain permanent water in small impoundments. Discussions in this chapter relate specifically to ponds such as those in semi-arid southwest and west-central Texas. Ponds in this area, as in most of the United States, were managed, if at all, specifically for recreational fisheries.

SPECIES COMPOSITION

Fish species inhabiting small impoundments are usually the result of stocking by man and invasion by species from either above or below the pond. Total number of species will be small in comparison with communities in large multi-purpose reservoirs.

In 20 permanent ranch ponds sampled over a six- to eight-year period in a three-county (Kinney, Maverick, and Webb) area in southwest Texas, number of species ranged from 1 to 12 (Table 4.1). Only two ponds contained over 10 species; both of these were located on large drainages with numerous ponds in the watershed above and a tendency to overflow for extended periods following heavy rains. The remaining 18 impoundments typically had small watersheds with no ponds above them. Mean number of species in these ponds was 5.7, with a range of 1 to 9. All ponds contained at least one game species; only five ponds were free of invading species.

In another three-year study of 21 ranch ponds in Zavala County, where topographic relief was considerably less, number of species was considerably higher (Table 4.1). Number of species ranged from 5 to 14, with a mean of 8.8 species per pond. All ponds again contained at least one game species. None was free of invading species. Maximum numbers of species (12 and 14) occurred in two impoundments with unusually large watersheds.

Twelve flood prevention lakes associated with floodwater-retarding structures in west-central Texas (Brown, Coleman, and Mills counties) had an intermediate diversity (Table 4.1), but otherwise were quite similar in species composition. Although these impoundments were much larger (mean area of 7.3 ha), they possessed certain features of construction that minimized chances of overflow (Noble et al. 1979), thereby limiting invasion

Table 4.1. Frequency of occurrence of fish species in south Texas ponds and west-central Texas floodwater-retarding structures (Common names follow Committee on Names of Fishes [1980])

Species	20 Upland Ponds	21 Low Relief Ponds	12 Floodwater-Retarding Structures
Largemouth bass	20	21	12
Bluegill	16	20	10
Redear sunfish	8	17	8
Hybrid sunfish[a]	2	1	0
Black crappie	1	4	0
White crappie	0	12	4
Channel catfish	6	9	11
Green sunfish	15	9	10
Longear sunfish	0	0	3
Warmouth	7	17	2
Spotted gar	1	7	0
Flathead catfish	0	0	2
Black bullhead	8	7	6
Yellow bullhead	1	2	1
Mosquitofish	6	2	2
Gizzard shad	9	17	1
Threadfin shad	3	2	0
Golden shiner	7	13	3
Carp	3	10	0
Red shiner	1	0	2
Notropis spp.	3	1	0
Fathead minnow	0	0	3
Smallmouth buffalo	2	12	0
Mexican tetra	3	0	0
Freshwater drum	1	0	0
Inland silversides	2	0	0
Rio Grande perch	0	1	0
Total taxa	22	20	16
Mean	6.2	8.8	7.2
Range	1–12	5–14	4–11

[a]Occurrence where parental species were not present, therefore likely stocked. Natural hybrids were encountered in most impoundments.

from downstream. Farquhar (1977) in a similar study of 56 flood prevention lakes in north-central Texas found an average of 7.9 species per impoundment. His study lakes, in a less arid area, had a mean surface area of 10.4 ha and differed markedly in species composition from those farther west.

Stocking of small impoundments typically is done to provide recreational fishing. Ponds smaller than 0.4 ha are recommended to be stocked with catfish (usually channel catfish), whereas larger ponds are stocked with largemouth bass (Texas Chapter, American Fisheries Society, 1980). Modde (1980) reviewed state stocking policies for small, warm-water impoundments and found that bluegill sunfish were most frequently stocked along with largemouth bass as a prey species, although much variation occurs among states, including those of the American Southwest.

Species are usually stocked in combinations which are likely to attain a balance (Swingle 1950) between predators (largemouth bass) and prey. A number of characteristics of balanced pond communities have been elaborated by Swingle (1950) and Anderson (1973). Basically, a balanced system will be a dynamic one, with continual reproduction by predator and prey species, diverse size composition of prey species so that food is available for all sizes of predators, and high growth rates for predators and prey. The balanced system will be productive, leading to satisfactory sustained yields of fish to the fishermen in proportion to the productive capacities of the pond. These criteria for balance were developed in the Southeast and Midwest, but are considered generally applicable to ponds which are managed for largemouth bass.

Pond management has been approached from the standpoint that simple systems are more manageable. This approach is manifested in the recommendation that small ponds are limited to channel catfish; a predator-prey system in a small pond would be too difficult to maintain in a balanced state if exploited. In bass ponds, the number of additional species is typically kept small, frequently just bluegills and perhaps also channel catfish. Redear sunfish, golden shiners, and threadfin shad are sometimes included as prey, but in addition to bluegills. Like bluegills and bass, redear sunfish occupy the littoral zone. As adults, they consume larger macroinvertebrates than do adult bluegills. In contrast, golden shiners and threadfin shad are open water species which feed principally on plankton. Catfishes, which are benthic feeders throughout most of their lives, consume principally insect larvae from the pond bottom. Although total standing crop of fishes tends to increase in proportion to number of species due to more complete use of resources (Bennett 1971), larger numbers of species complicate management by increasing complexity of the food web. The large numbers of species in ponds of southwest Texas (Table 4.1) would likely have similar effects.

Few ponds with only bass and bluegill were found in the study areas of southwest Texas for evaluation of the effect of species composition on standing crops. Biomass of largemouth bass in five ponds containing bluegills, other sunfishes, golden shiners, black bullheads or channel catfish, and gizzard shad was high ($\bar{x}$ = 99 kg/ha) in comparison with three ponds containing sunfish but no shad ($\bar{x}$ = 57 kg/ha). These standing crops of bass are much higher than those reported for unfertilized waters of the South, which are frequently around 30 kg/ha (Brenner 1983). However, the high estimate may be somewhat misleading since all were obtained at low water level, when most ponds had surface areas reduced by over 50% due to a two-year drought.

Although biomass of bass was high, these ponds were not necessarily in balance. Length composition of bass and sunfish seldom conformed to the desired Proportional Stock Densities (percentage of adult fishes of desirable size) suggested by Anderson (1980). Year class strength of bass varied markedly from year to year, and abundance of prey-sized fishes relative to bass was usually low compared to the suggested Y/C biomass ratios of Swingle (1950). During stable or declining water levels, growth rates of bass were frequently very low, particularly after the bass attained lengths of about 200 mm. In such cases, production (biomass elaboration) was negligible for adult bass.

ENVIRONMENTAL EFFECTS

Many environmental factors affect the productivity of small impoundments. Although these factors operate concurrently, three of the principal factors will be discussed individually as they pertain to ponds of semi-arid areas.

Water Level Fluctuation

Water level manipulation has long been considered to be an effective fishery management tool (Noble 1981). Lowering of water levels in the fall allows predators better access to prey. Raising of water level in the spring provides expanded habitat for reproduction and growth.

Of the physical factors affecting fish production in small impoundments of semi-arid regions, water level fluctuations probably have the greatest effects. Rainfall is not only limited, but it is also erratic. Furthermore, low humidity and high temperature accentuate evaporative losses. Ponds which have received no appreciable runoff in several years may suddenly be filled to overflowing. Accompanying this rapid rise in water level is a marked increase in surface area, commonly 100–200% or more. Certain species respond to the expanded habitat more rapidly than others, resulting in population explosions which may energize the remainder of the fish community or may lead to severe disequilibrium in the system. Largemouth bass appear to have a particular ability to produce a strong year class when water levels rise and new littoral habitat becomes available.

Declining water levels due to evaporation likely have some of the effects sought in artificial drawdowns of water level. Ponds and floodwater-retarding structures are often built with a small, deep, steep-sided borrow pit near the dam. During periods of drought, water level will recede into the borrow pit where the remaining water is less subject to evaporation due to decreased surface area and reduced exposure to wind. At this extreme, it is unlikely that any species will produce successful year classes. Further, the concentration of fish and absence of cover lead to prey depletion for all bass except those large enough (> 400 mm) to feed on abundant large prey. Rapid growth by these large bass seems to allow them to feed on progressively larger prey of a variety of species without becoming food-limited. Although such ponds will seldom go dry, some borrow pits will eventually become filled by sediment and some ponds are built without deep borrow pits. These will likely go dry during extended drought periods and subsequently become dominated by invading species unless stocked.

Extreme effects of water level fluctuations on largemouth bass were evident in two ponds which contained bass but no prey species. Following high water levels in spring 1976, populations were estimated by mark-recapture in fall 1977, after water levels had again fallen and surface areas were reduced approximately 75% and 50% in the two ponds. At that time, standing crops were 635 and 299 kg/ha respectively—far higher than other bass biomasses estimated. After one more year of low water levels, biomasses were 20 and 113 kg/ha respectively. The pond which showed the 75% reduction in area experienced a 97% decline in standing crop, whereas the pond which decreased 50% in area showed a 63% loss of bass standing crop. Intraspecific predation and starvation were responsible for these losses as the population was crowded into a reduced habitat.

Although no sequential biomass data were obtained on ponds with multiple species, sampling by electrofishing indicated that such marked changes in abundance were not typical. Nevertheless, major changes in species composition occurred from year to year, due at least in part to water level fluctuations. Water level fluctuations also have indirect impacts on fish production through effects on water clarity and vegetation. These will be discussed in the following section.

Water Clarity

Standing crops of fishes are expressed on a surface area basis (e.g., kg/ha) rather than on a volumetric basis, in part due to primary productivity being dependent upon sunlight. Plant growth, whether by phytoplankton or submergent macrophytes, occurs only in those surface waters with adequate light. Depth of light penetration depends upon suspended clay particles and shading from aquatic flora itself. Adequate water clarity is needed for photosynthesis and predation to occur efficiently, but too much clarity frequently leads to development of macrophytes and reduced ability of bass to catch their prey. The abundance of macrophytes also ties up nutrients, limiting phytoplanktonic productivity. Bushy pondweed (*Najas*) and muskgrass (*Chara*) seem particularly well adapted to hard waters of the Southwest and occur in most impoundments.

Small impoundments of semi-arid regions experience several unusual water clarity situations. Unlike large, mainstream reservoirs, watershed ponds are limited to areas where adequate clay is available for building the dam and sealing the bottom to prevent seepage. Ponds constructed in undisturbed watersheds should become clear after stabilization of the shoreline. (An exception to this is the pond with abundant clay-bottom shallows continually disturbed by wind action.) Unless substrates are unsuitable, submergent aquatic macrophytes will establish, thereby providing protection of the shoreline from wave action and enhancing sedimentation of clay particles. If water levels remain fairly constant, the macrophytes would grow to depths as deep as 2 m, thereby essentially occluding the littoral zone. As water levels slowly recede due to evaporation, the vegetation extends its distribution farther into the basin, maintaining its dominance of the littoral zone. At such time as water levels recede to the borrow area, the steep sides prevent much expansion of vegetation beds. Consequently, due to water level fluctuations, a pond may vary from having a large littoral zone with a high percentage of its area vegetated at moderate water levels to having a small littoral zone and a narrow band of submergents. Effects on fish habitat and prey availability differ markedly under these extremes.

At low water level, the effects of livestock watering become pronounced. Due to reduced shoreline, a greater proportion of bank is disturbed by a fixed number of livestock. Because cattle frequently wade a short distance into the water, they also trample much of the submergent vegetation. Consequently, it is not uncommon for the pond at low water levels to have reduced vegetation and increased clay turbidity. This reduced visibility for predation, if severe enough, will favor certain invading species such as the black bullhead (sometimes called mudcat) which further reduce water clarity by stirring up bottom sediments. Prolonged periods of drought may lead to a greatly reduced population of largemouth bass because of their reduced ability to feed, thereby favoring dominance by uncontrolled sunfishes, gizzard shad, and bullheads.

In ponds without borrow pits, the chronology is much the same, but mechanisms differ slightly. As water level becomes low enough for wave action to disturb the central basin of the pond, turbidity frequently increases. This is enhanced by livestock disturbances. Together these lead to minimization of submergents and to effects on fish population similar to those above. A heavy rainfall which leads to refilling of the pond may obliterate any remaining submergent vegetation stands. Along with increased water level, inflows of clay turbidity and nutrients (resulting in phytoplankton blooms) from the watershed tend to reduce water clarity, thereby shading the submerged weedbeds. The new littoral zone is briefly free of submergent aquatic macrophytes, but may include some inundated terrestrial vegetation which serves as fish habitat (Strange, Kittrell, and Broadbent 1982).

Water clarity has one additional effect of particular pertinence to pond fish communities—its impact on harvestability of largemouth bass. Under clear water conditions, particularly if submergent macrophytes are abundant, the bass will be hungry due to low availability of prey. They will frequently concentrate along the deep-water side of weedbeds in search of food. In addition to being hungry and concentrated, the bass can easily see lures in the clear water, thereby increasing their vulnerability to harvest. In such situations overfishing can easily occur, which can allow excessive increases in abundance of prey species which otherwise would be controlled by predation. In one clear, vegetated pond, 21% of the adult bass were caught in approximately one hour by several fishermen. In another, 22% were caught by two fishermen in two hours. In another even more extreme case, 50% were caught by one fisherman in two hours.

Fertility

Although application of fertilizers to pond waters can increase phytoplanktonic primary productivity and carrying capacity of fishes (Boyd 1979), this technique has little applicability to semi-arid region ponds. Naturally high carrying capacities apparently exist due to soil type, water quality characteristics such as high alkalinities, and nutrients available from livestock and other animal wastes in the watershed. Furthermore, because ponds infrequently overflow, nutrients are retained in the pond for extended periods of time between pond overflows. Biomass of gizzard shad in four study ponds ranged from 42 to 288 kg/ha, which suggests that primary productivity is high. When water levels are markedly reduced, extreme concentration of fishes occurs in the absence of artificial fertilization. This problem would be accentuated if fertilization programs were adopted.

In ponds with low productivity, which might benefit markedly from fertilization, several problems exist which make fertilization ineffective. Phosphorus, which is most likely to be limiting in old ponds, is apt to be precipitated in the hard waters common in the Southwest before it can be taken up by phytoplankton. If aquatic macrophytes are present, nutrients are more likely to lead to primary productivity of macrophytes than of phytoplankton, thereby tying up the nutrients instead of moving them through the food chain. Finally, effective fertilization programs involve continual applications, usually at intervals of two to three weeks, to sustain plankton blooms. Ranch ponds that are remote from main roads are not visited frequently; consequently special trips need to be made to apply fertilizers, thereby increasing costs of implementing such a program. Feeding of pelleted feeds with automatic or demand feeders to increase sunfish production might be a viable alternative for increasing total standing crop, but this technique has not been evaluated in the Southwest.

MANAGEMENT TO ENHANCE POND FISHERIES

Pond Construction

Although design criteria usually are based upon considerations other than fisheries, some aspects of pond design can minimize adverse effects on fish. Permanent ponds should be located singly on watersheds rather than in series or on major waterways. This will reduce flushing of ponds and invasion of undesired species during heavy rains. Ponds with gently sloping shorelines which vary much over 50% in surface area from full to borrow pit water level should be avoided. Such ponds will also be more subject to evaporative losses. Borrow pits should be constructed where possible. These not only serve as reserve water supplies for livestock but also as a refuge for fish during drought periods.

Elaborate designs have been developed for water discharge through standpipes and for tank watering of livestock below the dam rather than at the pond bank. These designs have little practical application to semi-arid areas. Because most watersheds are already well developed for water retention, relatively little can be accomplished through design modifications except when old ponds are replaced due to siltation or dam deterioration. Consequently, post-construction management holds greater potential for ponds of semi-arid areas.

Reclamation and Stocking

During periods of low water levels, it is frequently economical to remove all fish by treating the water with a toxicant, usually 1–3 ppm of 5% rotenone. (The higher rate will ensure that tolerant species such as black bullheads will also be killed.) Following detoxification (less than two weeks at summer temperatures), the pond can be restocked. Although reclamation can be done at minimal cost during low water levels, the long-term value of reclamation is often questionable. The undesirable species will frequently reinvade the first time major runoff occurs. With proper management, undesirable species may be held at manageable levels, thereby precluding the need for reclamation.

Although bluegills occurred in most ponds described in Table 4.1, absence of bluegills was encountered. In ponds that had been stocked with only largemouth bass (frequently these also contained green sunfish which had invaded) or that had been stocked with bass and hybrid sunfish (according to agency stocking recommendations in recent years), a severe prey deficiency always existed. These could be improved through introduction of small numbers of adult bluegills (Noble and Henry 1982) without reclamation.

Following filling of new ponds (or ponds that go dry) or reclamation of pond fish populations, stocking with desired species should be done immediately. Not only will such stocking allow rapid development of a fishable stock, but it will minimize chances of invading species establishing densities of problem size.

Stocking considerations involve selection of size of fish to be stocked as well as numbers, species, and times for stocking (Wenger 1972). Alternatives are usually limited if fish are obtained from public sources. In contrast, private producers are likely to have a variety of species and sizes available throughout most of the year. Split stocking chronology (rather than simultaneous stocking of all species) is frequently recommended to assist in survival and establishment of balance (Modde 1980), but this approach is time-consuming for small, remote ponds. When advanced fingerlings of bass, bluegills, and channel catfish are all available, a single late-winter or early-spring stocking according to the Texas recommendations (Texas Chapter, American Fisheries Society 1980), has been effective in establishing balance, since the bass and bluegills will spawn soon after stocking. This technique also reduces the time from stocking to fishing by nearly one year compared to that which occurs with typical fry stocking programs.

Habitat Management

Because of construction features, location, and unpredictability of runoff, small impoundments of semi-arid regions are not very conducive to habitat management. Two components of the habitat—clay turbidity and aquatic vegetation—can be feasibly managed to enhance production of desired species.

Turbidity can be minimized through both watershed and pond bank stabilization. Pond fisheries can't be the basis for setting land use practices, for example grazing systems, but small sedimentation ponds can be established in a watershed. Such structures have commonly been used to reduce soil erosion, but intensive development of sites in the watershed will reduce water supply to the principal impoundment while likely increasing overall evaporative losses.

Bank stabilization can be accomplished primarily through limitation of livestock access, particularly during low water levels. Extra-pond watering troughs, which have been designed to prevent livestock access, are seldom employed. Noble (1980) suggested partial fencing to limit livestock access to a few areas of a pond. This technique is more complicated for ponds which fluctuate markedly in water level. In most ponds, fencing to prevent access along one side with extension of the fence into the borrow area would be feasible. If built on the dam, this fence could simultaneously provide protection of the dam from erosion, thereby increasing the useful life of the pond. Periodic maintenance is required for fence sections which submerge and corrode.

Control of excessive submergent vegetation has been traditionally accomplished by chemical treatment. Chemical control has the attribute that, in many cases, localized control can be effected. Only a few chemicals are approved for use in aquatic vegetation control, and some have restrictions on livestock watering for short periods following application. Chemical control can be expensive since a single application seldom provides control for an entire growing season.

In recent years, biological control of submerged vegetation has been investigated. Since no native fishes are macrophyte consumers, exotic fishes have been used, primarily tilapias (especially *Tilapia zilli*) and grass carp (Shireman 1984). The tilapias, which are semi-tropical fishes, have had limited applicability because of overwinter mortalities. Grass carp have been widely used in ponds of the Southeast and Midwest but, like tilapia, are not legal for use in most other states including those of the Southwest, due to concern about their establishment and effects on non-target areas. Recent development of sterile triploid grass carp (Malone 1983) may lead to a more widespread availability of this fish for use in submergent vegetation control. Such control should be economical and feasible for remote ponds since long-term control (normally five years or more) can be anticipated. Although much research has been conducted on this species, its use has generally led to complete control of submergent macrophytes when stocked at recommended densities and sizes. Stocking strategies that allow limited control have not been developed, and no evaluations have been conducted in small impoundments similar to those discussed here. Although reduction in aquatic macrophytes would likely benefit fisheries of many small impoundments of the Southwest, elimination of vegetation stands would not be conducive to predator-prey relationships and would often increase chances of excessive turbidity.

Harvest Regulation

Because of susceptibility of largemouth bass to angling, harvest regulation is necessary to prevent their overfishing and to preserve balance in small impoundments. Conventional recommendations have been to remove 10 or more bluegills for every bass removed, but this recommendation is seldom followed because of the differences in catchability of bass and bluegills, the different fishing techniques used for them, and the differing appeal of the species to fishermen. Two alternative recommendations specific to largemouth bass have been developed. These are the use of annual harvest quotas and special size limits.

Annual harvest quotas can be established to prevent overharvest of largemouth bass if carrying capacity can be approximated. Because production likely exceeds carrying capacity in a dynamic, balanced community, a conservative annual quota would equal the bass biomass. This harvest would need to be spread over the year rather than being concentrated at any one time. Detailed record keeping is required for effective use of harvest quotas. This technique leads to high mortality rates of catchable-sized bass and favors development of populations comprised largely of young fish capable of rapid growth.

Until recently, minimum size limits have been used with limited success in small impoundments. Although minimum size limits help maintain balance by preventing overfishing of largemouth bass, their effect is to reduce bass size to or below the minimum limit, sometimes leading to overpopulation of small bass, which then leads to prey deficiencies and stunting of bass. Anderson (1976) proposed the slot limit (protected size range) as an alternative to minimum size limits. This approach prevents harvest of bass from an intermediate length range, frequently 30–38 cm, while harvesting fish from the stock which are smaller or larger than the slot size. This technique has particular application to small, remote impoundments since no record keeping is required, and probabilities of overfishing are minimized. With adequate levels of effort, slot limits should help prevent and overcome stunted bass populations in clear, densely vegetated ponds typical of many semi-arid areas.

Harvest is an integral part of developing and maintaining high productivity in small impoundments. Although standing crops of lightly fished ponds may be high, natural mortality of adult fishes is so low that growth rates are low under static conditions. Removal of fish, similar to expansion of habitat through increased water level, results in rapid production to fill the available niche.

ARTIFICIAL COMMUNITIES IN ARTIFICIAL IMPOUNDMENTS

Noble (1985) indicates that predator-prey relationships in large reservoirs are poorly developed because the reservoir community is artificial and natural equilibrium has not evolved. Early pond fisheries researchers identified a simple predator-prey combination which attained satisfactory balance in small impoundments of the Southeast and Midwest. Although numerous stocking combinations have been evaluated, the largemouth bass–bluegill combination has proven most satisfactory for warm-water ponds. This combination also is widely used in small impoundments of the semi-arid Southwest. These species principally use the littoral zone, with bluegills and small bass feeding on invertebrates and large bass feeding upon fishes. Additional species which use benthic and pelagic food resources should increase the biomass of fishes supported in a pond. Whether these additions would increase community stability in impoundments of semi-arid regions is uncertain. Severe fluctuations in water level, water clarity, plankton productivity, and submergent vegetation can lead to rapid changes in abundance of an individual species, thereby causing it to gain dominance or be reduced to negligible densities.

Management can be implemented to minimize the destabilizing factors that affect pond communities. New construction and structural modifications should take into consideration not only the need to hold water but to accommodate water level fluctuations without major changes in surface area or littoral area. After stocking appropriate fish species, harvest regulation can be both effective and economical. Sustained fishing, according to regulations that ensure that overharvest will not occur, will lead to productivity commensurate with basic fertility of the impoundment.

REFERENCES

Anderson, R. O. 1973. "Application of Theory and Research to Management of Warmwater Fish Populations." *Transactions of the American Fisheries Society*, 102:164–71.

———. 1976. "Management of Small Warm Water Impoundments." *Fisheries* (Bethesda) 1(6):5–7, 26–28.

———. 1980. "Proportional Stock Density (PSD) and Relative Weight (W_r): Interpretive Indices for Fish Populations and Communities." In *Practical Fisheries Management: More with Less in the 80's*, ed. S. Gloss and B. Schupp, 27–33. Ithaca: New York Chapter, American Fisheries Society.

Bennett, G. W. 1971. *Management of Lakes and Ponds*. 2d ed. New York: Van Nostrand Reinhold Co.

Boyd, C. E. 1979. *Water Quality in Warmwater Fish Ponds*. Auburn, Ala.: Agricultural Experiment Station, Auburn University.

Brenner, B. C. 1983. *Variations in Largemouth Bass Abundance and Standing Crop in Floodwater Retarding Structures in Central Texas*. Ph.D. diss. Texas A & M University, College Station.

Chapman, D. W. 1967. "Production in Fish Populations." In *The Biological Basis of Freshwater Fish Production*, ed., S. D. Gerking, 3–80. New York: John Wiley & Sons, Inc.

Committee on Names of Fishes. 1980. *A List of Common and Scientific Names of Fishes from the United States and Canada*. 4th ed. American Fisheries Society, Special Publication No. 12. N.p.

Farquhar, B. W. 1977. *Fish Population Characteristics of Flood Prevention Lakes*. Master's thesis, Texas A & M University, College Station.

Malone, J. 1983. "Malone's Sterile White Amur." *Farm Pond Harvest* 17(4):19–20.

Modde, T. 1980. "State Stocking Policies for Small Warmwater Impoundments." *Fisheries* (Bethesda) 5(5):13–17.

Noble, R. L. 1980. "Management of Lakes, Ponds, and Reservoirs." In *Fisheries Management*, ed., R. T. Lackey and L. A. Nielsen, 265–95. New York: John Wiley & Sons, Inc.

———. 1981. "Management of Forage Fishes in Impoundments of the Southern United States." *Transactions of the American Fisheries Society* 110:738–50.

———. 1985. "Predator-Prey Interactions in Reservoir Communities." In *Reservoir Fisheries Management: Strategies for the 80's*, ed., G. E. Hall, in press. Bethesda, Md.: American Fisheries Society.

Noble, R. L., and V. M. Henry. 1982. "Corrective Stocking of Forage-Deficit Ponds Using Bluegills." *Annual Proceedings of Texas Chapter, American Fisheries Society*, 5:54–65.

Noble, R. L., F. H. Sprague, W. C. Hobaugh, and D. W. Steinback. 1979. *Wildlife Benefits Through Construction and Management of Floodwater Retarding Structures*. Rocky Mountain Forest and Range Experiment Station General Technical Report 65:181–85.

Shireman, J. V. 1984. "Control of Aquatic Weeds with Exotic Fishes." In *Distribution, Biology, and Management of Exotic Fishes*, ed. W. R. Courtenay and J. R. Stauffer, 302–12. Baltimore: John Hopkins University Press.

Strange, R. J., W. B. Kittrell, and T. D. Broadbent. 1982. "Effects of Seeding Reservoir Fluctuation Zones on Young-of-the-Year Black Bass and Associated Species." *North American Journal of Fisheries Management* 2:307–15.

Swingle, H.S. 1950. *Relationships and Dynamics of Balanced and Unbalanced Fish Populations.* Alabama Polytechnical Institute, Agricultural Experiment Station, Bulletin 274.

Texas Chapter, American Fisheries Society. 1980. "Stocking and Management Recommendations for Texas Farm Ponds.

Wenger, A. 1972. *A Review of the Literature Concerning Largemouth Bass Stocking Techniques.* Texas Parks and Wildlife Department Technical Series No. 14.

5

Development of High pH in Mountain Lakes of the Southwest

Charles D. Ziebell and David E. Jones

SUMMARY

Elevated levels of pH associated with the development of various types of aquatic plants, including macrophytes and algae, have resulted in fish mortality in trout fisheries of the southwestern United States. A study of the pH levels of three lakes differing morphometrically, but typical of those in the mountains of Arizona, found that pH levels were always high during summer in dense canopies of rooted plants in the three lakes. However, the influence of plants on the pH levels was minimized where the ratio of macrophytes to open water was low. Because trout mortality due to pH shock has been high when fish are placed in shallow, high-level pH water, the authors have recommended releasing the fish in open water away from macrophytic beds where pH levels are lower. In addition, it has been suggested that pH be measured frequently, developing algae blooms be watched, weed harvesting be explored, and fish stocking be temporarily discontinued when pH levels exceed 8.9.

INTRODUCTION

In contrast to the low pH conditions caused by acid rain in the northeastern United States, high pH between 9.0 and 10.6 is a problem in numerous mountain lakes of the Southwest. These high pH values make it very difficult to manage the recreational trout fisheries because they often limit stocking during the summer when these lakes receive the greatest amount of fishing pressure. Occasionally high pH will cause fish kills which are economically and aesthetically undesirable.

High pH is an indirect result of natural or man-made activity in which excessive nutrients enter lakes or streams. More directly, high pH is primarily caused by the removal of carbon dioxide by aquatic plants. During photosynthesis plants can remove carbon dioxide faster than it can be replaced by respiration or diffusion from the atmosphere. This causes the pH to rise because carbon dioxide behaves as an acid in water and removal causes an equilibrium shift from hydrogen ions to hydroxl ions (Boyd 1979). But the level to which pH can rise in a given lake is also determined by the cation constituency of the alkalinity and the amount of aquatic plants in the area. The highest pH level that can be produced by carbon dioxide extraction in most lentic systems is near 10.0 when calcium carbonate is the predominant component of alkalinity (Boyd 1979). Increases in carbonate ion concentration caused by CO_2 removal by plants can eventually exceed the relatively low solubility product of calcium carbonate and result in precipitation. This tends to limit hydroxyl ion production and thus limits further increases in pH (Sawyer and McCarty 1978). However, intense photosynthetic activity in dense aquatic plant beds can result in pH exceeding 10.0 if calcium levels are low and alkalinity is comprised mostly of magnesium, sodium, or potassium carbonates which have higher solubility products and impose less stringent limits on hydroxyl ion production.

Lakes that have large, rapidly growing plant communities commonly produce pH levels well over 9.0. When rainbow trout are stocked as catchables for a "put and take" fishery, they come from hatcheries where the pH of the water usually is between 7.0 and 7.5. If the trout are stocked directly into lakes with a pH over 9.0, pH shock and fish kills can occur. Unfortunately, this happens in mid-summer when the fishing season is at its peak. State and federal wildlife agencies needed data to determine if there were ways to work around the problem. Our field studies focused around the temporal and spatial development of high pH in relation to the plant communities. Sampling procedures were designed to gain information on the potential impacts of plants on water quality and for implementation of stocking procedures that would avoid fish kills.

RESULTS AND DISCUSSION

The three lakes studied in Arizona between May and mid-September 1981 were different morphometrically and typified most of the lakes in the region. Rainbow Lake is a 32-ha impoundment with a maximum depth ranging from 5 m during the spring to 3 m in late summer. Aquatic macrophytes occupied the entire lake basin. Big Lake is a 194-ha impoundment with a maximum depth of 7 m. The shoreline slopes are rather steep, limiting the littoral zone, and macrophytes occupied only 6% of the lake. Concho Lake is a long, narrow, 53-ha impoundment that has a maximum depth of 4.5 m. Macrophytes were present in 35% of the lake basin. Each lake provided slightly different conditions and the pH varied depending on the size and depth of lake, the combination of macrophytes and algae, and the time of day.

Influence of Macrophytes

The macrophytes, or rooted aquatic plants, grew rapidly in early summer in the littoral zone of all three lakes. The size of these macrophyte beds and the thickness of the canopy (Fig. 5.1) determined the areas for the development of high pH. A comparison of pH levels within the aquatic plant beds of Rainbow Lake illustrates that the highest pH, above 10.5, occurred within the canopy at 0.2 m in the afternoon (Fig. 5.1). This rapid diurnal pH rise basically resulted from carbon dioxide removal during vigorous photosynthesis and low calcium levels which allowed an equilibrium shift to increase hydroxyl ions. Near the edge of the canopy about 20 m from shore, the increase was less and the highest pH value was 9.7. In samples taken beyond the canopy in open water but above the macrophytes, the pH level was 9.3. These very high pH values, which were restricted to water depths less than 0.5 m, occurred in the dense canopies of all three of the study lakes. In Rainbow Lake, macrophytes that grow from shore to shore play a very large role in the development of high pH throughout the entire lake.

In contrast to Rainbow Lake is Big Lake which has a very low percentage of shallow shoreline and rooted aquatic plants (Fig. 5.2). Submerged plants and canopies play a minor role in elevating pH of the lake because almost 95% of the area is open water. Macrophytes grew within restricted areas and produced high pH levels, but beyond the edges of the plants, pH levels dropped off rapidly (Fig. 5.3). The macrophytes clearly influence high pH, but if the percentage of the beds is low in relation to open water, the degree of influence is minimized. Consequently, the localized problem of high pH most likely does not have deleterious effects on the trout fishery.

Influence of Macrophytes and Algae

All lakes have individual characteristics which make them slightly different from one another. The morphometry of Concho Lake (long and narrow) was different from Rainbow Lake and Big Lake. The maximum width

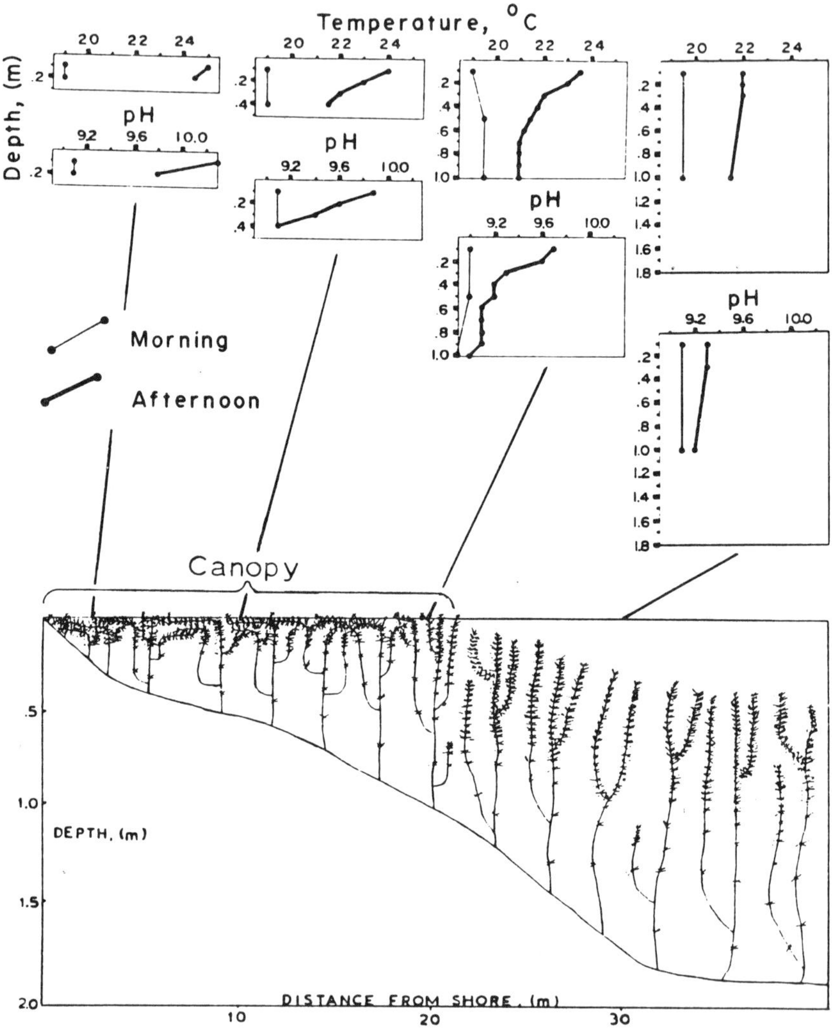

Figure 5.1. Morning and afternoon temperature and pH profile of a macrophyte bed in Rainbow Lake, Arizona on July 17, 1981.

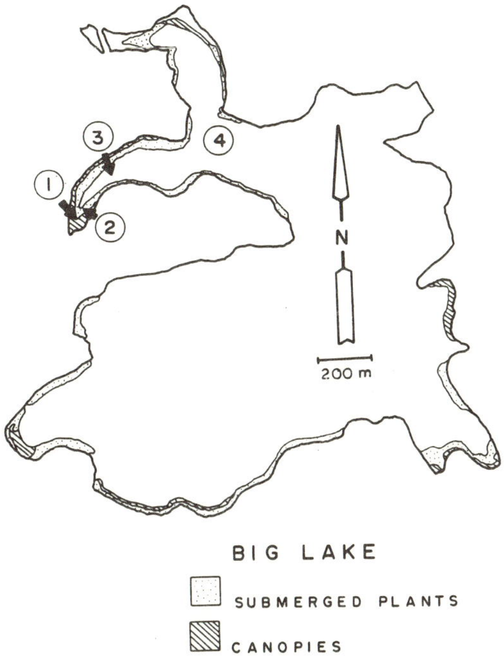

Figure 5.2. Big Lake, Arizona, with the locations of sampling stations, submerged macrophytes, and canopies.

was 460 m but approximately half of the lake was 200 m wide or less. About 35% of the lake contained macrophyte beds, and therefore the amount of limnetic or open water area was relatively small compared to Big Lake. The entire lake resembled a large cove and developed high pH values in two ways during two different time periods.

Data on pH from the canopy of macrophyte beds and from the limnetic zone illustrated that pH increases developed from different causes. In mid-June, the canopies of the aquatic plant beds produce pH values that ranged from 9.3 in the early morning to 10.3 in the afternoon. The range of Ph in that same area at 1 m was 9.3 to 9.4. Concurrently, pH values from the limnetic zone were similar to those at 1 m beneath the canopy (Table 5.1). In mid-July a profuse algae bloom occurred throughout the lake and pH values increased markedly to 9.8 at a depth of 2 m. The pH levels persisted through the first week of August before a gradual decline occurred. These data indicated that vigorous photosynthesis by macrophyte beds and by the phytoplankton in the limnetic zone combined to elevate the pH throughout

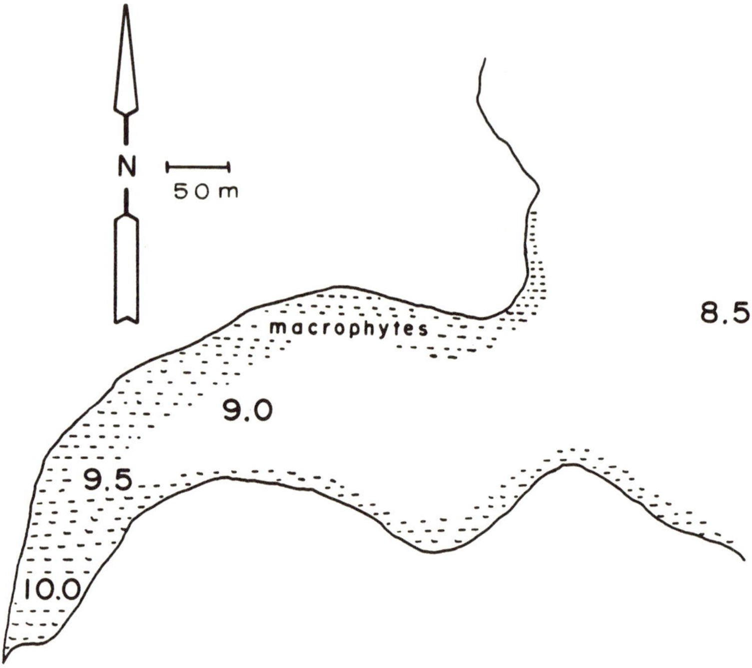

Figure 5.3. Value of pH inside and outside a cove in Big Lake, Arizona, on July 20, 1981.

the lake in the upper 2 m. The remaining volume of the lake with the pH in a more suitable range for fish was drastically reduced.

Variance in pH Development

A comparison of the data from the three study lakes illustrated a timing variance in the development of high pH. One common determinant for all lakes though was that plants and algae must be in a rapid growth phase to elevate pH.

In Rainbow Lake the pH rose throughout the spring resulting in a peak pH of 9.8 by mid-May. This level persisted until mid-June then steadily declined until the pH was below 9.0 by mid-July (Fig. 5.4). The drop in pH coincided with a change in the plant community. Starting in June, macrophyte beds consisting primarily of *Myriophyllum* began to thin out. Individual stems became sparsely foliated and heavily coated with epiphytic filamentous algae. Concurrently, *Polygonum* began to develop and by

Table 5.1. Selected data from Concho Lake comparing morning and afternoon pH values in macrophyte canopy and limnetic zones

		June 17		July 19		August	
Station	Depth (m)	AM	PM	AM	PM	AM	PM
Canopy	Surface	9.3	10.3	9.8	10.6	9.7	10.4
	0.5	—	—	9.8	9.6	9.7	9.6
	1.0	9.3	9.4	9.7	9.5	9.7	9.5
Limnetic	Surface	9.3	9.4	9.8	9.8	9.8	9.8
	1.0	9.4	9.3	—	9.6	9.8	9.8
	2.0	9.4	9.2	9.8	9.2	9.8	9.7
	3.0	9.2	9.1	8.8	8.1	—	—

mid-summer it was more conspicuous than *Myriophyllum*. Significant phytoplankton populations never developed.

Concho Lake had a pH of 9.0–9.3 from May to July at which time a rapid increase to 9.8 occurred. This coincided with the development of a profuse bloom of the algae *Rivularia* sp. and a sharp increase in chlorophyll-a (Fig. 5.4). A drop in pH did not occur when the *Rivularia* population diminished but pH did decline to 9.1 when chlorophyll-a decreased in September. A succession from *Myriophyllum* to *Polygonum* also occurred in Concho Lake.

In Big Lake the pH slowly rose throughout the summer until a pH of 9.6 was attained during the first week of September. This rise in pH was accompanied by an increase in the blue-green algae *Aphanizomenon* sp. and a rise in chlorophyll-a. An evaluation of the data from these three lakes clearly showed that each lake required a separate analysis of pH. Even though high pH development occurred in all lakes, the timing of development was asynchronous. The period of high pH was primarily controlled by the quantity and growth characteristics of the various types of aquatic flora. Because plant communities differ, the timing of pH elevation and decline is likely to be highly variable.

Fishery Management Implications

Mountain lakes in the Southwest provide many anglers with the opportunity to escape from the summer heat of the lower elevations and fish for trout in the cooler water. In recent years, fishing pressure on many of these lakes has increased to the level that stocking fingerlings in the fall to provide the fishing for the following summer is no longer sufficient. Consequently, state game and fish and federal agencies have adjusted manage-

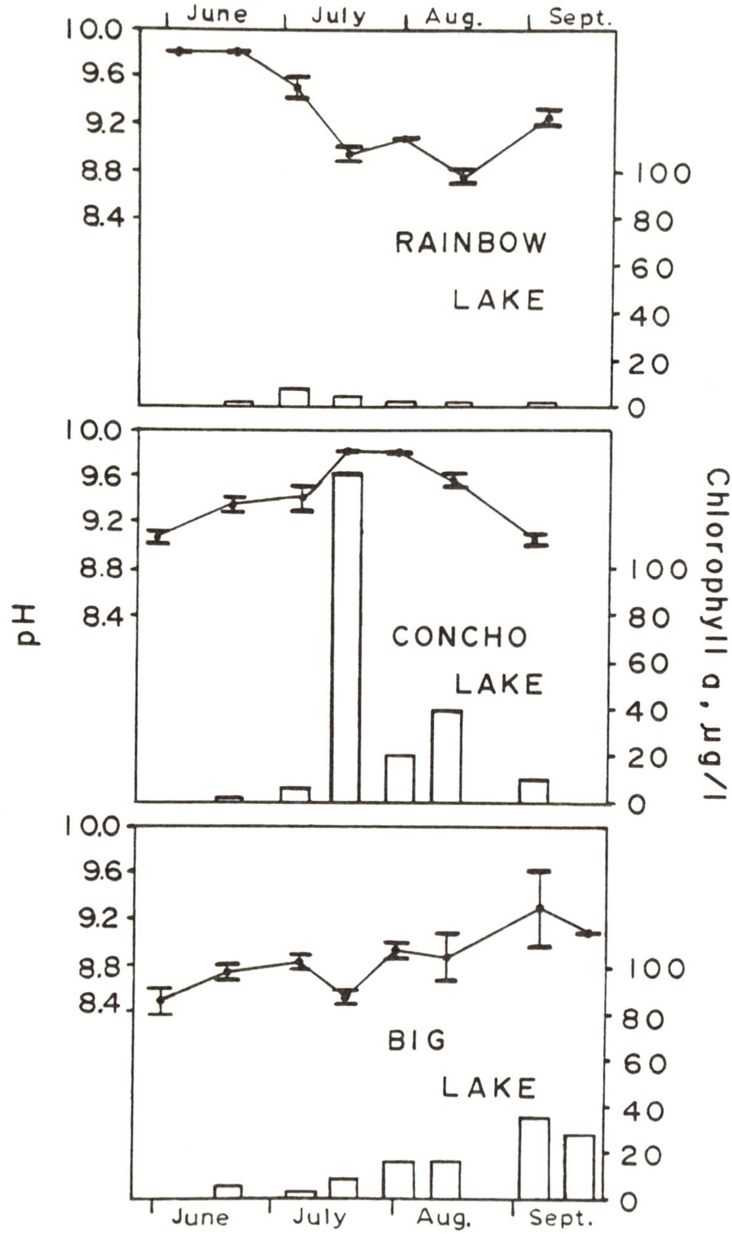

Figure 5.4. Diel pH range and mean (lines) and chlorophyll-a (bars) during the summer in Rainbow, Concho, and Big lakes, Arizona.

ment procedures to include frequent stocking of catchable-sized rainbow trout for "put and take" fisheries. During the peak of the summer fishing season, stocking rainbow trout involves a certain degree of risk because of high pH. The degree of risk and the direct harmful effects to the trout, however, are not well documented. Strong evidence now exists that trout subjected to pH values over 9.0 experience shock and at pH 9.5 25–50% mortality can be expected (Witschi and Ziebell 1979; Murray and Ziebell 1984). The problem of risk takes place when the trout are released from hatchery trucks into shallow water adjacent to large aquatic plant beds, and hence in areas with high pH. The effects of high pH are not evident immediately following stocking as mortalities do not occur until after several hours of exposure. Furthermore, many of the trout that die sink to the bottom of the lake and may not be seen again. If management procedures call for stocking every week, as is the case in many lakes, a large percentage of the fish released may never reach the angler's creel.

The data from this study can be used as a guideline to minimize mortality of stocked trout. Several factors must be considered because the vast range of vegetation characteristics in the mountain lakes results in a high degree of variability in the timing and duration of high pH. First, the pH should be measured frequently throughout the season near shore and out in the open waters at various depths to determine the areas that have the lowest pH values. These areas are most likely to be in open water at a depth of 2 m. Second, stocking should be avoided in or near macrophyte beds. If necessary, a means to transport the trout offshore into the open water should be provided. Although offshore stocking is laborious, it may be considerably safer. Third, surveillance of developing algae blooms in open water should be maintained. If pH levels exceed 8.9, stocking should be discontinued unless acclimation procedures such as those suggested by Murray and Ziebell (1984) are implemented.

Although these guidelines will assist in minimizing stocking mortality of rainbow trout, the basic problem of excessive aquatic plant growth and the associated high pH needs additional attention. One approach being taken by the Arizona Game and Fish Department is intermittent "weed harvesting" during the spring and summer months. That agency has purchased a harvesting machine and has removed plants in critical areas of some of the problem lakes. Some of the preliminary data have been encouraging. The removal of macrophytes has resulted in decreased pH levels within tolerable ranges for trout. The study will continue to determine if mechanical harvesting provides a means for controlling high pH in the mountain lakes.

If this method proves to be successful, it would certainly be beneficial to the summer "put and take" trout fisheries in the Southwest.

REFERENCES

Boyd, C. E. 1979. *Water Quality in Warm Water Fish Ponds*. Auburn University, Agricultural Experiment Station, Auburn, Alabama.

Murray, C. A., and C. D. Ziebell. 1984. "Acclimation of Rainbow Trout to High pH to Prevent Stocking Mortality in Summer." *The Progressive Fish-Culturist* 46(3):176–79.

Sawyer, C. N., and P. L. McCarty. 1978. *Chemistry for Environmental Engineering*. 3d ed. New York: McGraw-Hill.

Witschi, W. A., and C. D. Ziebell. 1979. "Evaluation of pH Shock on Hatchery Reared Rainbow Trout." *The Progressive Fish-Culturist* 41(1):3–5.

6

Ecology of Small Desert Playas

Steven J. Loring, William P. MacKay,
and Walter G. Whitford

SUMMARY

Desert playas, common in the southwestern United States, have not been intensively studied despite the fact that they are simple systems that can be easily manipulated to test ecological theories. Data from a playa located in New Mexico were compared to earlier data and those from other playas to identify the effects of flooding. Tests of hydrology and chemistry showed the following: water was lost to percolation and evaporation; potassium concentrations were higher than those for sodium; and neither the playa water nor soil had a significant increase in chemical compounds from the drainage area following flooding. Although new vegetative growth resulted, repeated flooding reduced perennial grass and herbaceous plant production. Three main life cycles (fast onset and immediate development; fast onset and prolonged development; and rapid development and delayed onset) were identified; microorganisms were used to define trophic levels of the most common animals, and a food web was constructed. Although organisms with quick development were unaffected, the unusually high population densities that followed reflooding (within three months) increased competition for food and predation among slow-growing animals. North American playas are similar to those in other parts of the world and have such physiological adaptations as large thermal differences between bottom and surface waters, rapidly maturing inhabitants, and low dissolved oxygen levels. Although insects, birds, and wind contribute, dispersal of temporary-pond species is not assured or common. To fully understand playa environments, broad-based comparisons of experimental data (for example, on the effects of land management practices or the importance of playas in recharging groundwater) are needed.

INTRODUCTION

Playas are desert basins with no surface drainage which become ephemeral freshwater ponds or lakes following rainfall. Size and duration can vary greatly, but water depth is usually less than one meter (Crawford 1981). Playas, also known as pans, chottes, and sebkas in other parts of the world (Crawford 1981), are extremely common in the southwestern United States (Reeves 1972). Eastern New Mexico and western Texas alone are estimated to have at least 20 to 30 thousand playas (Sublette and Sublette 1967). Some aspects of the physiognamy and geomorphology of playas are known (Neal 1975; Merickel and Wangbert 1981), but the biological and chemical dynamics in such systems generally have been ignored.

Fauna of a few grassland temporary ponds have been investigated (Prophet 1963; Horne 1967, 1974; Donald 1983), but comparable data from desert playas are largely lacking. The fauna and flora of desert playas are potentially interesting to many ecologists. Playas are sporadic aquatic systems that are inundated by large disturbance events such as heavy rain. The physiological adaptations which allow organisms to complete their life cycles in a few days and remain dormant in dry soils for years are fascinating. Yet playas are relatively simple systems that readily lend themselves to manipulation, allowing tests of ecological theory.

Playas in the arid regions of the southwestern United States are important to the cattle ranching industry. Many are like the Jornada playa: nonsaline with a cover of grasses (e.g., *Panicum obtusum*) palatable to cattle. Because the shallow playas hold water for a very brief time, most have been modified by the construction of earthen stock tanks that may hold water for several months. This allows cattle to use the large standing crops of grasses that develop after a flood. The consequence is that stock tanks hold water more frequently than do entire playas and remain as aquatic habitats long after the playas dry.

It is important to have a better understanding of playas. Many of them are used by mineral collectors and by the general public as playgrounds for off-road vehicles. Others are located on military reservations where they are used for many purposes, including weapon testing. They are prime candidates for missile installations and for hazardous chemical or nuclear waste repositories. Playas are important in replenishing aquifers in desert regions. Therefore, playas must be studied and their ecological roles better understood before these uses are continued.

We reviewed studies of a playa located on the New Mexico State University Ranch which is part of the Jornada Long-Term Ecological Research (LTER) study area. We compared our data with earlier studies conducted as part of the U.S. International Biological Program (IBP) Desert Biome Project and with data from other playas.

The alluvial fan outwash areas (*bajada*) drain from the slopes of Mt. Summerford into a 12-ha playa basin (Fig. 6.1). The slope varies from 10%

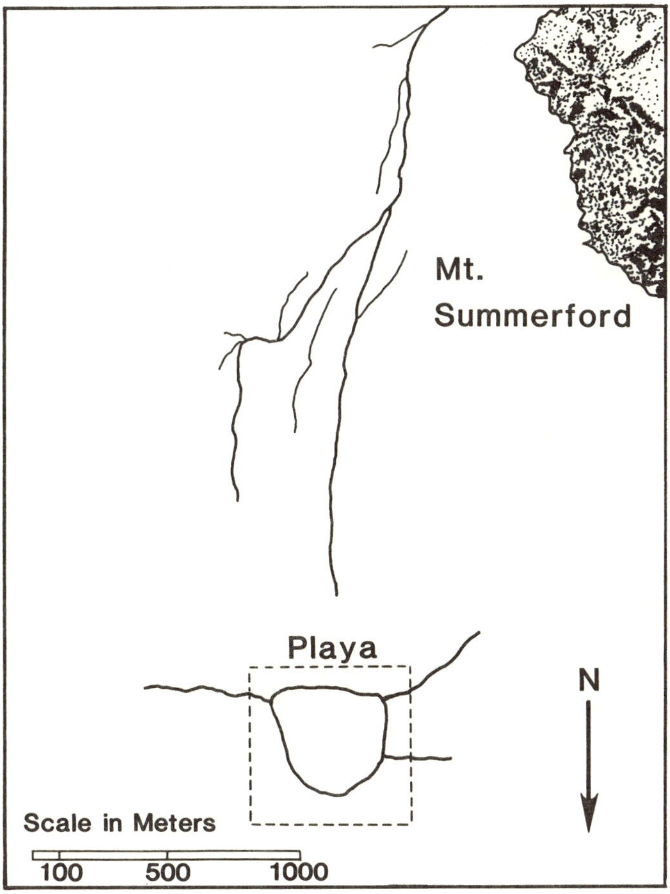

Figure 6.1. Relationship of the Jornada playa and the Mt. Summerford bajada.

on the bajada to less than 2% on the playa fringe. The playa has a maximum depth of 140 cm (Fig. 6.2) and has been undisturbed by humans, except for the excavation of an open-ended earthen stock tank (< 5% of the playa surface area) with a depth of 90 cm. Soils vary from coarse gravels on the outwash areas to heavy clay silts on the playa bottom. Following rains, the playa soils swell, causing accumulation of runoff water.

Flooding normally occurs sporadically during summer once every two to four years. In order for a desert playa to flood, it must receive a large volume of runoff in a short period of time or a large quantity of rainfall continuously within four days after the playa soils are saturated and sealed. Of the recorded events that have led to flooding of the Jornada playa, all but one have been intense storms of short duration that produced runoff from the watershed. The Jornada LTER playa flooded once in 1970, three times in 1972, twice in 1974, only partially in 1977 and 1980, and twice in 1984 (May and August). Before it completely dried after the August 1984 flood, the playa was reflooded twice.

We examined the climatic history for the Jornada Experimental Range and Las Cruces area and counted the number of months since 1914 that had total rainfall of 50 mm or more. This figure was chosen because it takes about 50 mm of rainfall during one storm for the playa to flood. Although

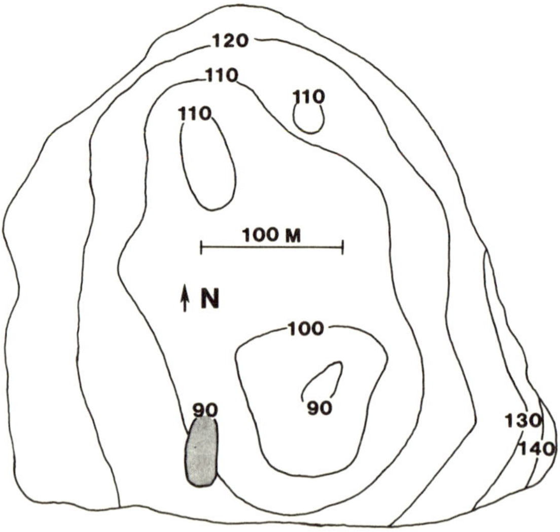

Figure 6.2. Topographic map of the Jornada playa. Units are in cm. The darkened area represents the cattle tank.

not all months with total rainfall of at least 50 mm will have had storms of sufficient intensity to flood the playa, counting them allows us to make a rough estimate of the likelihood of flooding during any one month. The probability of 50 mm or more precipitation per month is .34 in July, .39 in August, .46 in September and October, but only .03 in May. The storm in May 1984 produced 49 mm in 2.5 hours on the watershed. This was the most rainfall for a single storm in May since records have been kept.

HYDROLOGY AND CHEMISTRY OF THE PLAYA

One of the most important features of the Jornada LTER playa is that water is lost by percolation to deep soil layers in addition to losses by evaporation. During dry periods, the soil-water potentials at 90 cm may be nearly equal to those at the surface (Fig. 6.3). When the playa flooded during the IBP program, the soil-water potential of the upper 90 cm of soil was at less than −0.03 mPa (Fig. 6.3). In 1970, when the playa was flooded only in mid-summer, the soils in the upper 90 cm began to dry soon after standing water disappeared. In 1972, flooding in autumn kept playa soils saturated until the following April (upper 15 cm) and May (90 cm). Small rain events during 1971 and the first half of 1972 effectively wet only the upper 15 cm. Although the data on soil moisture are only for the upper 90 cm, these data coupled with evaporation data suggest that there is percolation from the flooded playa into deep soil layers.

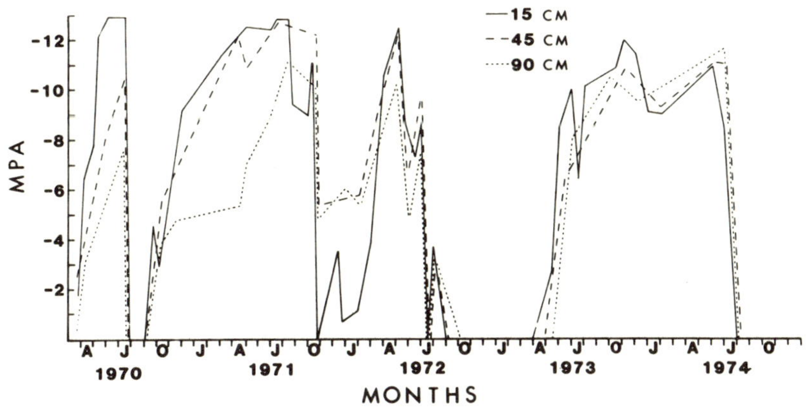

Figure 6.3. Soil-water potentials at three separate depths on the bottom of the playa, 1970 to 1974.

Water loss in the spring 1984 flood occurred gradually until the playa was dry after about two weeks and the tank was dry after about 60 days (Fig. 6.4). The second series of floods in 1984 resulted in two major recharges of the system after the initial flood. Water was present in the tank for over one hundred days (Fig. 6.4). Water loss could be due to evaporation (including evapotranspiration), infiltration, or both. One would expect evaporation to be the most important factor as the environmental temperatures were very high, and infiltration to be relatively unimportant due to the heavy clay soils of the playa. Evaporation normally would result in the concentration of ions, especially sodium and chloride; however, these ions showed no concentration effect (Fig. 6.5), suggesting that evaporation was relatively unimportant in the loss of water. Therefore, infiltration and recharge of the water table appeared to be the most important factors. Apparently, playas play an important role in maintaining desert aquifers.

Another surprise was the ratio of sodium and potassium in the playa water (Fig. 6.6). Normally, sodium concentration is higher than potassium concentration (Wetzel 1975). In the Jornada LTER playa, potassium concentrations were about an order of magnitude larger than sodium concentrations. This is extremely unusual.

An unusual correlation between water chemistry and biological activity was observed following the May 1984 flood (Fig. 6.6). Alkalinity and concentrations of calcium and magnesium increased initially but dropped sharply about Day 35 after the flood. This corresponds with a decline in

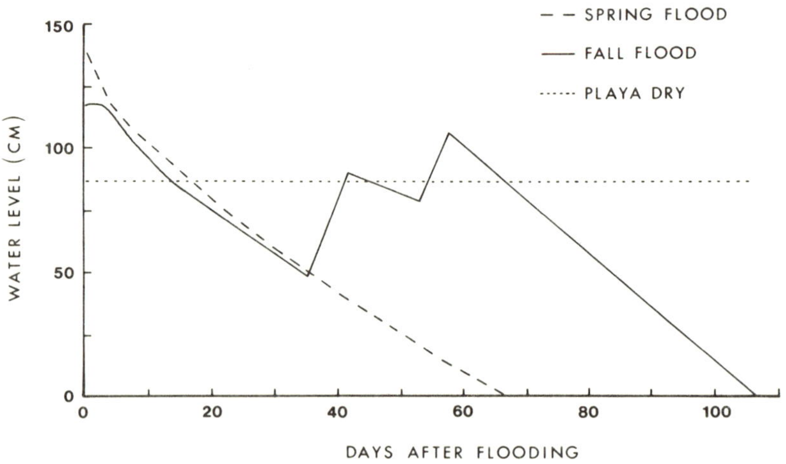

Figure 6.4. The water level in the playa and tank after the two major floodings of 1984.

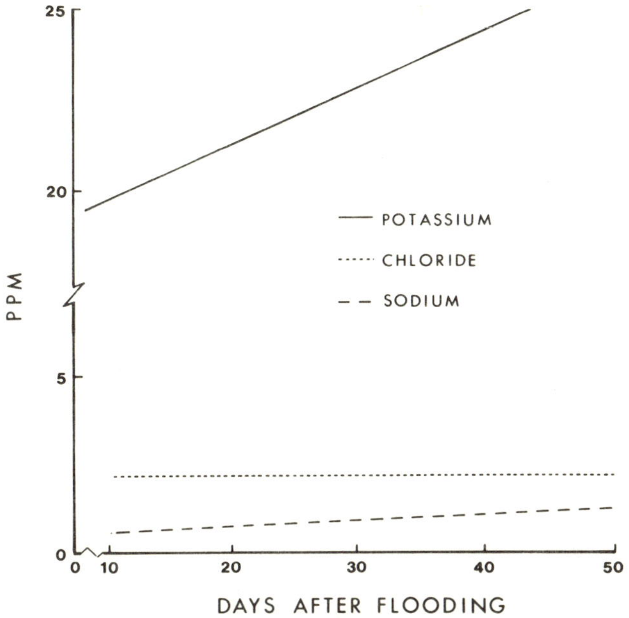

Figure 6.5. Changes in concentrations of potassium, chloride, and sodium ions over time. The curves are composites of the 1970 flood and the two floods of 1984.

population densities of water fleas and the fairy shrimp, *Streptocephalus texanus*. This apparent correlation may be purely coincidental.

Because playa basins receive water runoff and organic matter from the bajada, we originally expected the playa to act as a nutrient sink for the entire drainage area. However, neither the playa waters nor the soils had significant increases in chemical compounds (Figs. 6.5 and 6.6). The question becomes one of where and how the leached chemicals are transported. The Jornada LTER playa does not accumulate salts (Fig. 6.5) as do many playas in the Great Basin. Many chemicals (salts, for example) may percolate through the soil with the water until they reach the water table. Once they enter the desert aquifer, they may be transported. A more immediate vehicle of nutrient transport out of the playa is in the form of toads. Tadpoles must store enormous amounts of nutrients to enable them to survive the following dry spell as toads. Most toads burrow into the soils on the playa, but many toads leave the playa basin before burrowing into the soil for aestivation. Some toads have been found at surprising distances

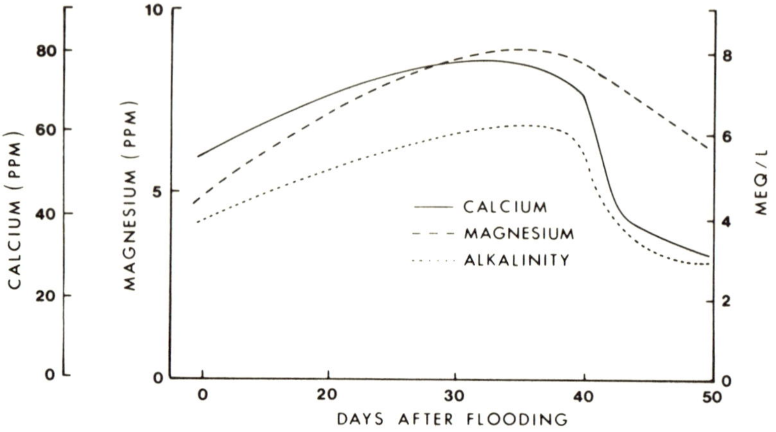

Figure 6.6. Changes in concentrations of calcium and magnesium ions over time. Data are from the spring 1984 flood. Alkalinity change over time is also indicated.

(1–2 km) upslope from the playa. The emigration of toads could represent a considerable loss of nutrients from the playa basin.

VEGETATIVE RESPONSE TO PLAYA FLOODING

The flooding cycles affect the productivity patterns of a playa. The perennial grass *Panicum obtusum* responds with rapid vegetative production when the soils are wet to 15 cm (Fig. 6.3). Flooding results in death of above ground parts. Additional new growth may follow a flooding event as occurred in 1970 (Fig. 6.7). Repeated flooding of a playa reduces productivity of the perennial grasses and of other herbaceous plants (Whitford 1973, 1975). The high biomass production on a playa bottom contrasts with the grass production on the surrounding watershed (1.09 kg · ha^{-1} in 1970, 12.2 kg ha^{-1} in 1971, and 0.64 kg · ha^{-1} in 1972), while on the playa bottom, production of *P. obtusum* was 1100 kg · ha^{-1} in 1970, 300 kg · ha^{-1} in 1971 and 100 kg · ha^{-1} in 1972 (Fig. 6.6). These data underscore the importance of playas for domestic livestock, not only as a source of water but also of large-quantity, high-quality forage.

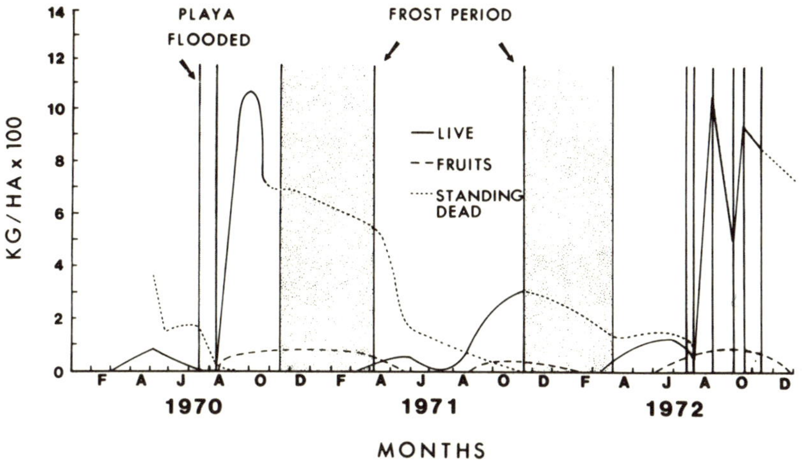

Figure 6.7. Responses to flooding by *Panicum obtusum* on the playa, 1970 to 1972.

ANIMAL RESPONSES TO FLOODING

Noy-Meir (1973) suggested that pulse-reserve models effectively explain much of the desert physical and biological responses to discrete pulses of rain. According to this model, a pulse of production should be triggered by rain. Although much of the production is lost by consumption or mortality, some is diverted into a reserve, such as seeds. Studies at the Jornada LTER site have shown that many desert organisms do not respond to rain as predicted by the pulse-reserve model (Whitford et al. 1983). On the other hand, playas should support the pulse-reserve model. Organisms such as fairy shrimp, tadpole shrimp, and toads must have a large pulse of water to fill the playa, which allows them to store food resources to survive the lengthy, intervening dry period.

Even though playa organisms should respond according to a general pulse-reserve model, the normal response times of various organisms vary considerably and fall into three main life cycle responses: (1) fast onset, immediate development; (2) fast onset, prolonged development; and (3) rapid development with delayed onset (Fig. 6.8). The first life cycle behavior was found in mosquito larvae and pupae (*Aedes* sp.), clam shrimp (*Eulimnadia texana* [Conchostraca]), and tadpole shrimp (*Triops longicaudatus* [Notostraca]). These animal species can complete their life cycles within two weeks, often less. This allows them to develop before competi-

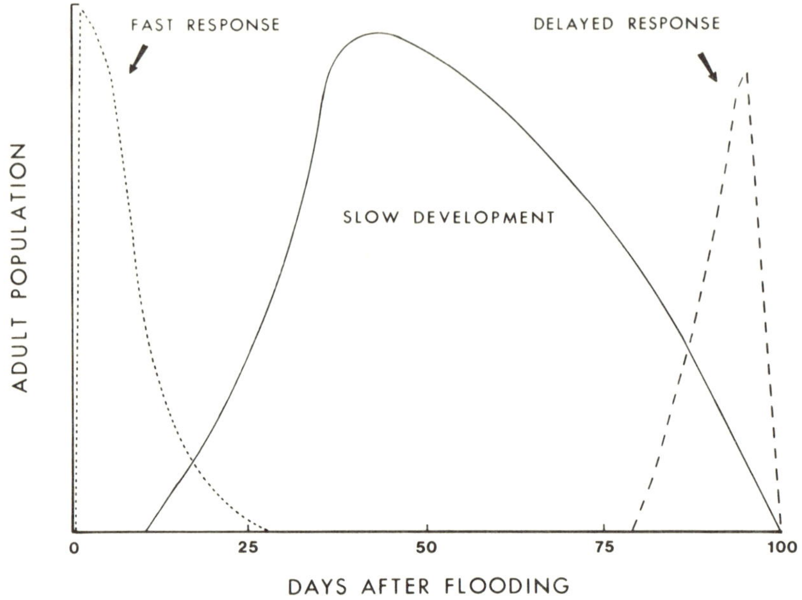

Figure 6.8. Three general life cycle responses that occur in desert playa animals.

tion for resources or, possibly, predation from slower-developing animals becomes severe. An analogy would be the early blooming of spring flowers in the forest understory before the tree canopy closes and blocks most light.

The long-term behavior of slower development is used by water fleas (*Moina wierzejskii*), fairy shrimp (*Streptocephalus texanus* and *Thamnocephalus platyurus*), and tadpoles of spadefoot toads (*Scaphiopus multiplicatus*). Each of these organisms grow, use, and store resources as long as possible before laying eggs (Crustacea) or leaving the water (toads). *Scaphiopus* tadpoles develop into predatory and herbivorous morphs after they get large.

During a study of breeding biology and survival of recently metamorphosed toads in the July-August 1974 flood, Creusere and Whitford (1976) estimated the numbers of breeding adult anurans (calling males, gravid females) at 3,000 *Scaphiopus multiplicatus* (*hammondi*), 2,500 *Scaphiopus bombifrons*, 1,000 *Scaphiopus couchi*, 500 *Bufo debilis*, and 150 *Bufo cognatus*. The previous year (1973) when only the stock tank in the playa held water, the breeding population consisted of 700 *S. multiplicatus*, 200 *S. bombifrons*, and less than 20 *S. couchi*. In the 1984 studies, we noted the presence of sexually mature adults of these species but found only

Scaphiopus multiplicatus tadpoles in the playa until 50 days after the initial flooding when *B. cognatus* tadpoles first appeared.

Creusere and Whitford (1976) also provided data on overwinter survivorship of spadefoot toads (Pelobatidae) and true toads (Bufonidae) juveniles. An initial population of 20,000 bufonid juveniles in late summer 1972 was reduced to 150 by mid-summer 1974. An estimated total of 200,000 juvenile *Scaphiopus* dropped to 1,800 after two years. These data showed that the survivorship of sub-adult anurans followed a negative exponential curve. The survivorship of juvenile anurans is a function of their stored lipids (Whitford and Meltzer 1976), and the survival of newly metamorphosed desert anurans is undoubtedly a function of the body mass of the tadpole at the time of transformation. Larger juvenile toads can dig into the soil and survive the mid-day conditions of the drying or dry playa (Creusere and Whitford 1976). Smaller individuals seek shelter under surface objects and many of these succumb to the severe environment.

Using the data in Creusere and Whitford (1976) and Whitford and Meltzer (1976), it is possible to estimate the translocation of nutrients from the playa to the surrounding areas by juvenile toads that die in burrows during the first winter at 445 kg fresh weight of anuran biomass or 62.4 kg dry mass.

The third strategy of delayed onset of development was shown by *Bufo cognatus* tadpoles during the May 1984 flood. These tadpoles were not numerous until Day 50 of the flood. At this time, when population levels of fairy shrimp and *Scaphiopus* tadpoles were low, the *B. cognatus* tadpole population rapidly increased and remained high until the water finally dried. Presumably this strategy avoided competition with the Crustacea and other tadpoles and possibly also avoided predation from *Scaphiopus* tadpoles.

Most of these species passed through a single generation during any one flooding. An exception was the water flea population (Fig. 6.9). The spring population went through three populations during a period of 50 days, shortly before the tank dried (Fig. 6.4). The fall 1984 population may have also passed through three generations. The first generation corresponded to the first generation of the spring flood. The fall population did not show a second generation of adults until 60–70 days after the playa flooded. The third generation was smaller with densities similar to the first flooding, and occurred after 80 days. The considerable length of time (50 days) between the first and second generation in the fall flood is especially interesting. During the same length of time there were two separate generations in the spring flood. This postponement may have been due to the effect of predators or to competition, possibly by rotifers of the genus *Brachionus*.

The population densities of rotifers following the fall 1984 flood demonstrated an interesting chronological sequence (Fig. 6.10). No rotifers were

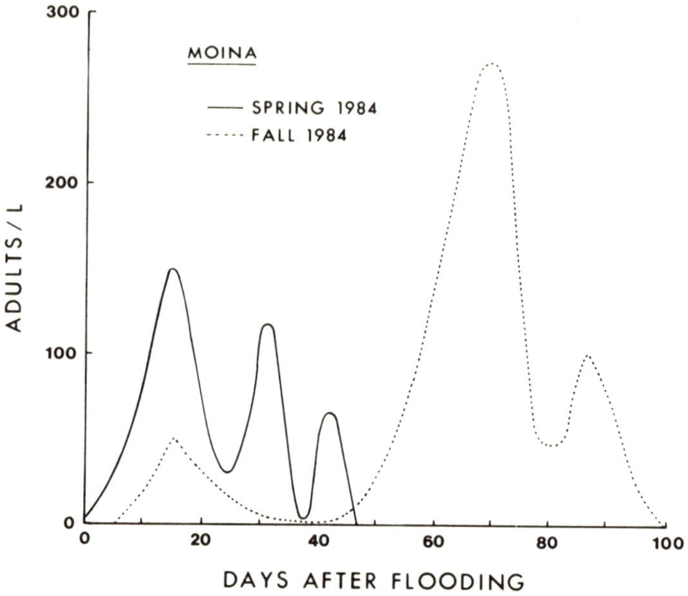

Figure 6.9. Changes in the water flea adults (> 1.0 mm body length) (*Moina wierzejskii*) populations in the spring and fall 1984 floodings.

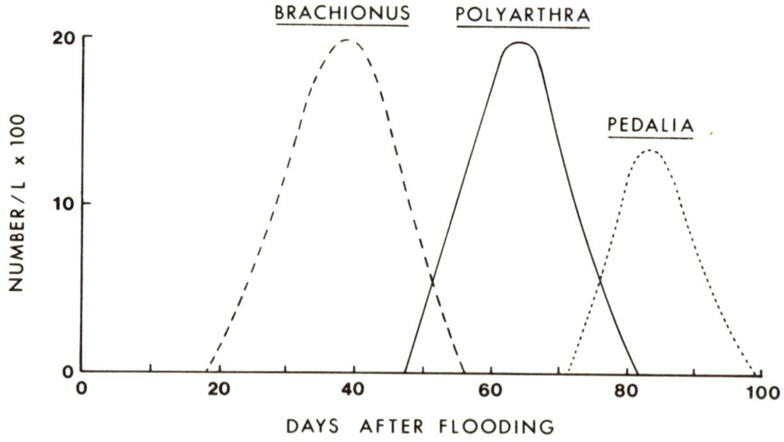

Figure 6.10. Population densities of three genera of rotifers in the Jornada playa. *Brachionus* curve is a composite of the spring and fall floods. *Polyarthra* and *Pedalia* were found only in the fall flood of 1984.

common until 20 days after initial flooding. *Brachionus* spp. were the first to reach high population densities after about 40 days. This genus was present in both the spring and fall floods. After the densities of *Brachionus* dropped, the population of *Polyarthra* reached high densities (at 65 days). Later, when population densities were again low, the population of *Pedalia* increased. Other rotifers were present, but their population densities were very low. *Polyarthra* and *Pedalia* were not collected during the spring flood, possibly because the water was not present for a long enough period of time.

Crustacea found in ephemeral desert ponds characteristically tolerate low dissolved oxygen levels (Belk and Cole 1975). We measured changes in oxygen levels and water temperatures in the Jornada playa (Fig. 6.11). Oxygen levels and water temperatures at the bottom showed little daily change but there were considerable diurnal fluctuations in these parameters at the surface. Lowest oxygen levels occurred during the early morning when oxygen demand by poikilotherms was lowest due to the low temperatures. Later in the day, both the water temperature and oxygen levels increased from solar radiation and photosynthesis by algae and water plants, respectively. Surface temperature and oxygen level dropped during the night to reach low levels again by morning.

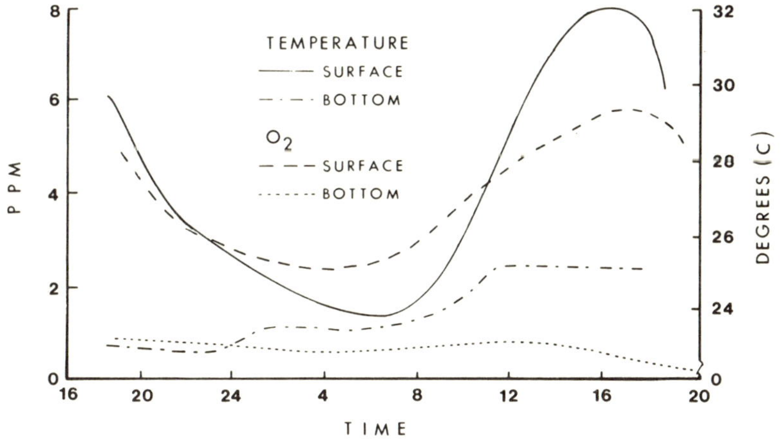

Figure 6.11. Daily changes in dissolved oxygen and temperatures in surface and bottom waters of the Jornada desert playa in 1970.

INTERSPECIFIC TROPHIC INTERACTIONS

We used microcosms in the tank to define the trophic levels of the most common animals in the playa. Both species of fairy shrimp are filter-feeders. Tadpole shrimp are primarily scavengers and do not prey on the fairy shrimp, although they have well developed mandibles. *Schaphiopus* tadpoles are known predators of other tadpoles (Bragg 1964; Woodward 1983), and we hypothesized that they also may prey upon fairy shrimp and tadpole shrimp. We found that the tadpoles did not eat tadpole shrimp, but were important faculative predators of fairy shrimp. Laboratory observations demonstrated that tadpoles would not actively hunt shrimp, but would capture and eat shrimp whenever they encountered one. The predaceous tadpole morph would attack other tadpoles but, because the tadpoles were all large, they were unable to do more than remove pieces from the tails.

We used the data to construct a food web (Fig. 6.12). The playa begins

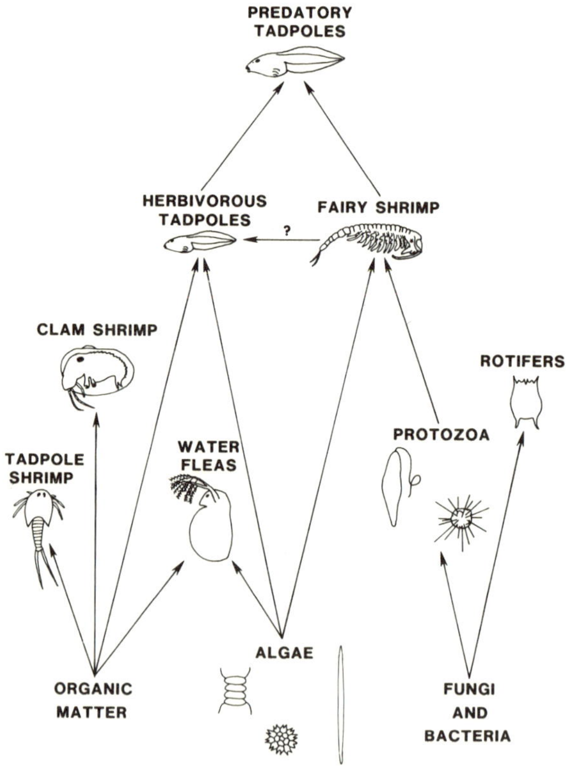

Figure 6.12. Food web of organisms in a desert playa.

as a heterotrophic system based on dead organic matter washed into the playa. Initially, the nitrate levels are very low but by Day 10 the level peaks because of organic matter breakdown (Fig. 6.13). As the playa water shifts from a heterotrophic system to an autotrophic system, the nitrate levels drop. Within a few days, the tadpole shrimp, bacterial, and fungal populations peak, and other populations (such as protozoans, rotifers, water fleas, and fairy shrimp) begin to increase. Diatoms, other algae, and rooted vascular plants begin to reach high densities, supporting the water fleas, fairy shrimp, and herbivorous tadpoles. Predaceous tadpoles appear after a few weeks and the playa food web is completely formed (Fig. 6.12).

We were able to determine the effect of abiotic and biotic influences (especially competition and predation) as a result both of our microcosm studies and the effects of two floods during 1984. This year was unusual in that the playa flooded again within one and a half months of its becoming dry after the first flood. Ordinarily, years pass between playa floods and natural mortality culls the aestivating toads and resistant eggs of Crustacea and rotifers. Because the intervening time was very short, nearly all eggs and organisms probably survived. Therefore, initial populations of organisms were extremely high after the August 1984 flood. This was very interesting because we were able to observe how unusually high population densities affected the life cycle responses.

Those organisms which normally used the quick development mode were unaffected by high initial population densities; mosquito larvae, tadpole shrimp, and clam shrimp still developed and disappeared quickly. Among more slowly developing animals, however, competition for food resources,

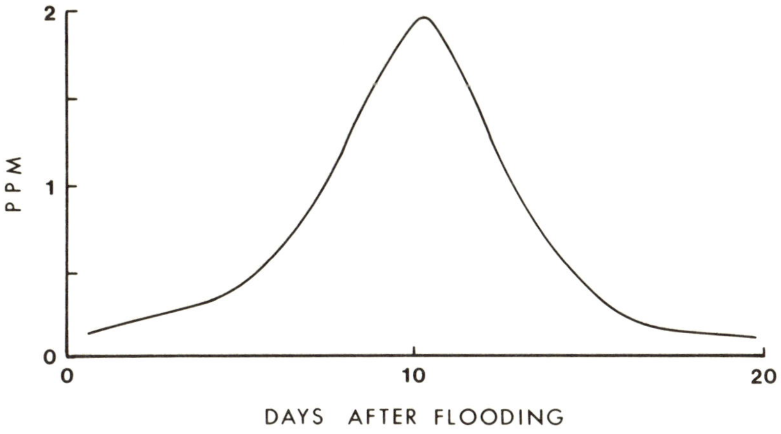

Figure 6.13. Nitrate levels after the 1970 flood.

if not predation, appeared intense. Populations of water fleas and fairy shrimp rapidly decreased to extremely low levels, presumably because of competition and predation from *Scaphiopus* tadpoles, and did not begin recovery until *Scaphiopus* tadpole density was reduced by metamorphosis of tadpoles into juvenile toads. In fact, fairy shrimp populations never did increase again. The physical size of *Scaphiopus* tadpoles was smaller during the second flood; tadpoles did not become large enough to develop into different morphs, as they did during the first flood. This means that the stored resources in each tadpole are smaller than normal and the probability of survival is much lower (Creusere and Whitford 1976; Whitford and Meltzer 1976). Most of the tadpoles from the late summer and fall flood probably would not survive until the next flood. This hypothesis could be tested by estimating the size of the one-year-old juvenile population at the beginning of the next summer. It was apparent that much of the rapid decline of the large fairy shrimp populations after the August 1984 flood was due to predation from large numbers of *Scaphiopus* tadpoles because the chance of interspecific contact was much higher than normal. Tadpoles are not predators of tadpole shrimp, but, because the tadpole shrimp develop more quickly than the tadpoles, there probably is not much potential for interaction between these groups under natural conditions.

Our data suggest that in normal years competition for food resources (suspended organic matter and algae) may be more important than predation in regulating population densities. However, when initial populations are extremely high, as happened after the August 1984 flood, predation appears at least as important as competition in explaining initial population fluctuations. This is because facultative predators (like *Scaphiopus* tadpoles) have more opportunity to eat shrimp.

At Taylor Well, a semi-permanent playa located on the Jornada Experimental Range, the vertebrate predator is the salamander *Ambystoma tigrinum*. This playa was studied in the late 1960s by Whitford and Becker (1972). *Ambystoma tigrinum* apparently filled the same functional ecological role at Taylor Well as did tadpoles at the Jornada LTER playa. Larval and neotenic adult salamanders were filter-feeders of bacteria, algae, and microcrustacea. Non-neotenic adults fed on aquatic insects and fairy shrimp (*Thamnocephalus platyurus* and *Streptocephalus texanus*), but when crowding occurred, they became cannibalistic. The population densities of fairy shrimp at Taylor Well, as at the Jornada LTER playa, showed a remarkable drop (134/m^3 on July 4, 1969, to 12.7/m^3 by July 29, 1969) which was correlated with the growth of salamanders.

DECOMPOSITION

Decomposition processes in the playa also were examined (Fig. 6.14). Playas receive a large pulse of organic matter from the drainage areas and decomposer organisms must act rapidly during the brief period of standing water. A large fraction of litter mass loss in water is from leaching (Saunders 1976; Carpenter, Odum, and Mills 1983). Weight loss occurs quickly, often within the first 24 hours, and may be as high as 27% (Peterson and Cummins 1974; Briggs and Maher 1983; Carpenter, Odum, and Mills 1983).

Fungi and bacteria develop on the leaf material and begin the process of decomposition. Later, invertebrates graze on the fungi and process the litter. We have no large invertebrate shredders in our playa. We evaluated the importance of predators of fungi and bacteria in May 1984 by using litterbags with two mesh sizes. The larger mesh allowed most invertebrates to enter the bags, while the small mesh allowed access only to very small invertebrates. When we excluded most invertebrates, the mass loss rates for creosotebush (*Larrea*) litter were higher than for those bags with large invertebrates present (Fig. 6.14); therefore, playa macroinvertebrates act

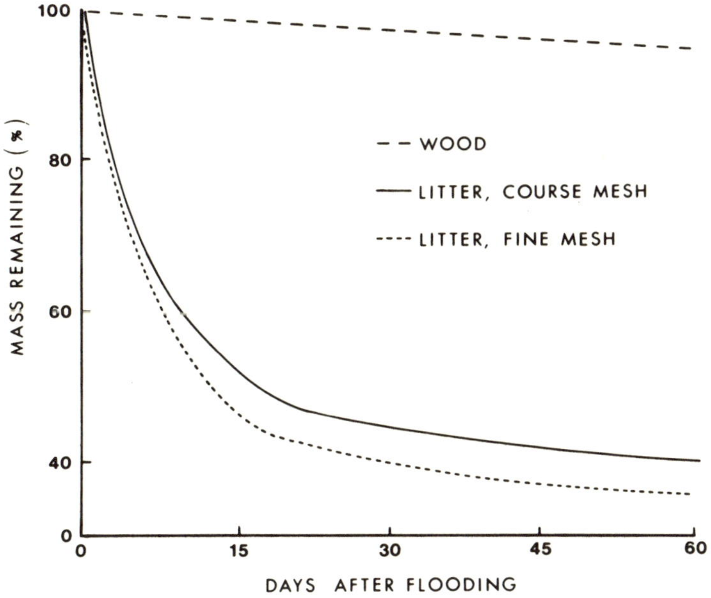

Figure 6.14. Decomposition of creosote wood and leaf litter in litter bags of two mesh sizes, spring flood 1984.

as regulators of the creosote litter decomposer community. No decomposition was detected in 1974 when *Panicum*, filter paper, and creosotebush leaves and twigs were suspended in the flooded playa. The conclusion then was that the flooding period was too short for decomposer organisms in the water to colonize and appreciably decompose the substrates.

HETEROTROPHIC OR AUTOTROPHIC SYSTEM?

In 1970, a series of abiotic measurements were made on the flooded playa (Whitford and Ludwig 1971). The water was nearly neutral, pH 7.1–7.0, until 8–10 days after flooding at which time the pH gradually increased to 7.3–7.5 There was a peak in nitrate nitrogen eight days after flooding followed by a steep decline in NO_3-nitrogen (Fig. 6.13). The increase in NO_3-nitrogen was paralleled by an increase in dissolved potassium. An algal bloom was noted during the period of rapid decline in NO_3-nitrogen. The diel fluctuations in dissolved oxygen on August 6–7, 1970, are what would be expected with an algal bloom (Fig. 6.11). However, in 1972 when the playa flooded O'Brien (in Whitford 1973) reported on studies using $^{14}CO_2$ in light and dark bottles and incorporation of labelled ^{3}H thymidine[2] and ^{14}C glucose to estimate overall biosynthetic activity. During the period when the playa held water in July and in September–October 1972, there was no evidence of an algal bloom, no differences in $^{14}CO_2$ incorporation in light and dark bottles, and no differences in thymide and glucose incorporation. O'Brien concluded that the playa was a heterotrophic system because there was no evidence of autotrophic activity. Although these were limited data, they do underscore the differences in the biological activity in this playa in the 1970 flooding and in the 1972 flooding. Before 1970, the playa had been heavily grazed and when the playa flooded in 1970, there was little organic matter on the playa bottom. When the playa flooded in 1972, there was 500 $kg \cdot ha^{-1}$ of litter and standing dead *Panicum obtusum* and 30 $kg \cdot ha^{-1}$ of dead herbs (primarily *Hymenoxys odorata*). This large mass of dead plant material probably provided sufficient substrate to maintain the system as a heterotrophic system. In the 1972 flooding, heterotrophic activity apparently was supported by organic material transported into the playa from the watershed. If heterotrophs are not energy limited, they will use nutrients in the production of more microbial biomass as rapidly as nutrients are released by metabolic activity, which will maintain the heterotrophic nature of the system. However, if there is a lack of carbon source, the dying microbes will release nutrients that can support the growth of autotrophic algae. These relationships are reasonable explanations for the apparent differences of primary production in this playa when comparing the 1970 inundation with that of 1972.

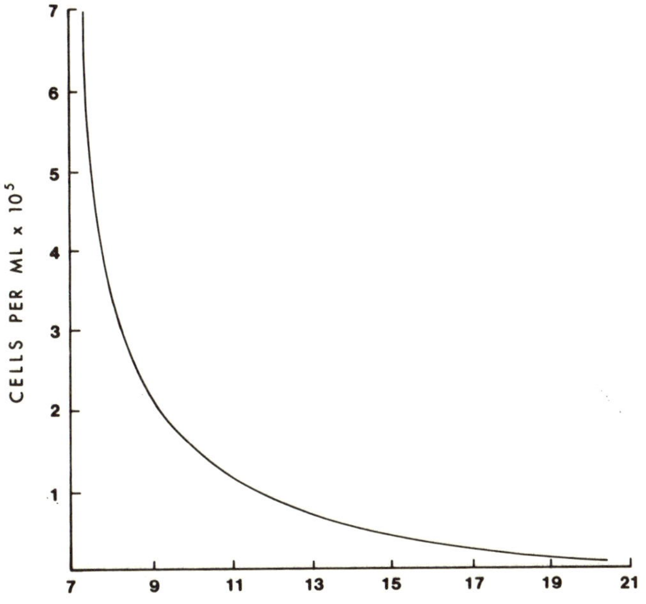

Figure 6.15. Bacterial population densities in the Jornada playa, 1974.

The playa remained heterotrophic when it flooded in 1974. Even so, the counts of bacteria dropped rapidly after high initial levels (Fig. 6.15). This is significant when one realizes that playa aquatic microflora may be of considerable importance as a nutrient source for both vertebrates and invertebrates (Whitford 1973) (Table 6.1).

Table 6.1. Use of bacteria by aquatic animals during the 1972 flooding of the Jornada LTER playa

Animal	Initial Cpm[a]	Final Cpm	% Used	Cpm/ Animal
Tadpole	11,803	4,432	63	10,411
Fairy shrimp	30,399	20,396	33	6,500
Tadpole shrimp	34,089	20,475	41	15,500

[a]Bacterial cells were labeled with ^{14}C glucose. Cpm represents relative density of bacterial cells added.

COMPARISON OF TEMPORARY PONDS

North American playas are similar to desert playas in other parts of the world. The crustacean composition of North African temporary ponds often consists of the same, or closely related, genera as are found in North America (Rzóska 1961, 1984). This is not surprising given that the physiological and environmental requirements are the same, and that playa fauna are almost entirely restricted to temporary ponds. In addition, most of these crustacean groups are widespread and of ancient lineage.

Ephemeral waters are present on every continent. In temperate climates, temporary waters (e.g., vernal ponds) usually last much longer than in more arid climates. By their temporary nature, however, they impose a strong selective force on the life in them. It is hardly surprising, therefore, to find the same families of organisms inhabiting temperate and tropical temporary ponds (Rzóska 1961; Pennak 1978).

Aside from relative duration, the major difference between desert and temperate ephemeral waters may be one of predictability. In temperate climates, vernal or autumnal ponds may fill regularly every year, while desert playas fill sporadically. Even so, the physiological problems shared by aquatic organisms facing prolonged dry periods are similar. The season, duration, and temperature of flooding was important in determining the dominant invertebrates in some temperate temporary ponds (Prophet 1963; Donald 1983).

PHYSIOLOGICAL ADAPTATIONS

Temporary ponds often contain several different temperature regimes. According to Eriksen (1966) there can be a 7°C temperature difference between bottom and surface waters of shallow pools. *Triops granarius* (tadpole shrimp) and various species of *Streptocephalus* (fairy shrimp), both desert phyllopod crustaceans, are able to choose the most suitable temperature available in the playas (Cloudsley-Thompson 1966).

Ephemeral ponds require that their inhabitants mature rapidly. Larvae of *Triops longicaudatus* (tadpole shrimp) and *Thamnocephalus platyurus* (fairy shrimp) were found by Hillyard and Vinegar (1972) to have extremely high Q_{10} values for oxygen consumption (4.12 and 11.81, respectively). These authors speculated that growth rate is also increased when water temperature is high.

Diel cycles of oxygen concentration, often dropping to very low levels, characterize both desert and non-desert temporary ponds (Moore and Burn 1968; Horne 1971; Hartland-Rowe 1972). Physiologically, phyllopod crustacea can tolerate low dissolved oxygen concentrations. For example, *Triops longicaudatus* (fairy shrimp) and *Streptocephalus texanus* (fairy shrimp), which are both found in the Jornada LTER playa, had LD_{50} values of 2 hours in 20–30°C water with 0.55 and 0.65 ppm dissolved oxygen (Horne 1971). Some species, such as *T. longicaudatus,* also have special blood pigments (e.g., hemoglobin) which may aid their tolerance to low dissolved oxygen concentrations (Horne and Beyanbach 1971). Phyllopod crustaceans also employ behavioral adaptations to low oxygen levels. In both desert and mesic temporary ponds, fairy shrimp, tadpole shrimp, and clam shrimp will swim to the surface and fan the water (Moore and Burn 1968; Horne 1971, Hillyard and Vinegar 1972). We saw this behavior in the Jornada LTER playa, especially during the hot part of the day when oxygen demands were high. These adaptations are not limited to desert playas, but generally characterize phyllopod crustacea of temporary waters.

DISPERSAL

Temporary ponds are isolated, so their inhabitants must have various means of dispersal. Passive dispersal of eggs and resting stages by insects and birds is probably the most important method (Maguire 1963). Arnold (1966) established that crustacean eggs could be carried on the feet of wading birds. Proctor, Malone, and DeVlaming (1967) found that killdeer (*Charadrius vociferus*) ingest crustacean eggs, many of which remain viable after passing through their guts, and contribute to the dispersal of crustacea. Hartland-Rowe (1972) speculated that wind may play a part in dispersing crustacean resting eggs, but had no data as to its relative importance.

The dispersal of temporary-pond inhabitants is neither assured nor common. Sometimes temporary ponds which are nearby do not even share the same fauna (Bishop 1967). Frey (1971) and Belk (1974, cited by Belk and Cole 1975) found considerable morphologic variation between crustacean populations from different ephemeral ponds, indicating that gene flow and dispersal between ponds are uncommon. Donald (1983) concluded that colonization probably was unimportant in temporary ponds near Calgary, Alberta.

RESEARCH NEEDS

Desert ephemeral lakes (playas) with clay-silt bottoms that seal incompletely are characterized by water infiltration into groundwater stores. In such playas, there is negligible salt accumulation in the standing water and in the surface soil materials. The lack of any significant salinity allows the development of an ephemeral lake flora and fauna not unlike that found in ephemeral waters in more mesic environments. As the limited data from the studies of playas on the Jornada have shown, the energy base may vary depending upon the land management practices imposed on the playa. Without grazing, the large standing stock of organic matter in a recently flooded playa provides energy for heterotrophic organisms (bacteria and possibly some fungi). If organic material is sparse, dissolved nutrients will be available to support growth of phytoplankton and the energy base will be autotrophic. Input-output budgets are needed for desert playas. Which sources of organic matter are most important as determinants of metabolic characterization of the playa—autochthonous or allochthonous? How do allochthonous inputs vary with storm intensity and duration? Do inputs vary as a function of antecedent conditions on the watershed? More data are required in order to develop a real understanding of changes in playa water chemistry and quantity.

Apparently, there is little difference in the grazing trophic level in heterotrophic and autotrophic systems. The first consumer level in both systems is made up of filter-feeders, e.g., clam shrimp, fairy shrimp, tadpole shrimp, rotifers, and small tadpoles.

The available data show that successful playa organisms have variable life history patterns, all of which are beneficial under certain circumstances. The data also confirm that the population dynamics and phenology of a flooded playa can vary considerably depending upon antecedent conditions and inputs from a given flood event. There are obviously differences in playas which are modified to provide livestock with water for long periods compared with those having stock tanks designed to hold water for only one to two months after a flood event.

Playas and stock tanks are important features of semi-arid grassland and desert rangelands. The hydrological characteristics of these features need considerable further study. It would be useful to be able to predict the infiltration-percolation losses from a flooded playa on the basis of playa-bottom soil characteristics. This would allow assessment as to the importance of playas in groundwater recharge over a broad geographical area. It is also important to know how the chemistry of playas varies depending upon sources of inputs. For example, the Jornada playa has low sodium levels and higher potassium levels. The watershed alluvia are from monzonitic granite rocks of the Dona Ana Range. Is the water chemistry of a playa in large part a function of the basic geology of the watershed?

There are many questions about the ecology of desert playas that require broad-based comparative and experimental data. For example, what are the implications of length of drying time and flooding periodicity to the life cycles of playa organisms? Are the initial processes, flora, and fauna the same in playas (or tanks) that flood annually as in playas that flood irregularly? Some playas have been so modified that they hold water nearly all year. The dry period in these may be a few weeks and the tank may not dry at all in some years. How do these compare with the truly ephemeral playas? As we have indicated in this chapter, our knowledge of ecosystem processes in flooded playas is fragmentary and, in many instances, anecdotal. Carefully designed studies using experimental techniques are required to test the validity of many of these ideas and to provide a basis for understanding the role of playas in desert landscapes.

ACKNOWLEDGMENTS

We thank many people for laboratory and field assistance, especially David G. Peterson, Solange Silva, Jill Each, Enrique Lessa, Danforth Weems, and David Lightfoot. Drs. Thomas M. Frost and Carl Bowser assisted in interpretation of the results, and Dr. Marsha Conley helped with data analysis. Dr. Robert O'Brien allowed us to include unpublished data. Brenda Massad typed the manuscript. The investigation was supported by NSF DEB-8114466 (Long Term Ecological Research Grant) and NSF BSR-8406708.

REFERENCES

Arnold, G. P. 1966. "Observations on *Lepidurus acticus* (Pallas) (Crustacea, Notostraca) in East Greenland." *Annual Magazine of Natural History* (Series 13) 9:599–617.

Belk, D. 1974. *Zoogeography of the Arizona Anostraca with a Key to the North American Species*. Ph.D. diss., Arizona State University, Tempe.

Belk, D., and G. A. Cole. 1975. "Adaptational Biology of Desert Temporary Pond Inhabitants. In *Environmental Physiology of Desert Organisms*, ed. N. F. Hadley, 207–26. Stroudsburg, Pa.: Dowden, Hutchinson & Ross.

Bishop, J. A. 1967. "Some Adaptations of *Limnadia stanleyana* King (Crustacea: Branchiopoda: Conchostraca) to a Temporary Freshwater Environment." *Journal of Animal Ecology* 36:599–609.

Bragg, A. N. 1964. "Further Study of Predation and Cannibalism in Spadefoot Toads." *Herpetologica* 20:17–24.

Briggs, S. V., and M. T. Maher. 1983. "Leaf Fall and Leaf Decomposition in a *Juncus effusus* Population." *American Wildlife Naturalist* 86:28–45.

Carpenter, J., W. E. Odum, and A. Mills. 1983. "Leaf Litter Decomposition in a Reservoir Affected by Acid Mine Drainage." *Oikos* 41:165–72.

Cloudsley-Thompson, J. L. 1966. "Orientation Responses of *Triops granarius* (Lucas) (Branchiopoda: Notostraca) and *Streptocephalus* spp. (Branchiopoda: Anostraca)." *Hydrobiologia* 27:33–38.

Crawford, C. S. 1981. *Biology of Desert Invertebrates*. New York: Springer-Verlag.

Creusere, F. M., and W. G. Whitford. 1976. "Ecological Relationships in a Desert Anuran Community." *Herpetologica* 32:7–18.

Donald, D. B. 1983. "Erratic Occurrence of Anostracans in a Temporary Pond: Colonization and Extinction or Adapation to Variations in Annual Weather?" *Canadian Journal of Zoology* 61:1492–98.

Eriksen, C. H. 1966. "Diurnal Limnology of Two Highly Turbid Puddles." *Verhand lungen der International Vereinigung fur Theoretische und Angewandte Limnologie* 16:507–14.

Frey, D. G. 1971. "Worldwide Distribution and Ecology of *Eurycerus* and *Saycia* (Cladocera)." *Limnologiue Oceanographer* 16:254–308.

Hartland-Rowe, R. 1972. "The Limnology of Temporary Waters and the Ecology of Euphyllopoda." In *Essays in Hydrobiology*, ed. R. B. Clark and R. J. Wooten, 15–31. Exeter, England: University of Exeter Press.

Hillyard, S. D., and A. Vinegar. 1972. "Respiration and Thermal Tolerance of the Phyllopod Crustacea *Triops longicaudatus* and *Thamnocephalus platyurus* Inhabiting Desert Ephemeral Ponds." *Physiological Zoology* 45:189–95.

Horne, F. 1967. "Effects of Physical-Chemical Factors on the Distribution and Occurrence of Some Southeastern Wyoming Phyllopods." *Ecology* 48:472–77.

———. 1971. "Some Effects of Temperature and Oxygen Concentration on Phyllopod Ecology." *Ecology* 52:343–47.

———. 1974. "Phyllopods of Some Southern High Plains Saline Playas." *Southwestern Naturalist* 18:475–79.

Horne, F., and K. W. Beyanbach. 1971. "Physiological Properties of Hemoglobin in the Branchiopod Crustacean Triops." *American Journal pf Physiology* 220:1875–81.

Maguire, G., Jr. 1963. "The Passive Dispersal of Small Aquatic Organisms and Their Colonization of Isolated Bodies of Water." *Ecology Monograph* 33:161–85.

Merickel, F. W., and J. K. Wangbert. 1981. "Species Composition and Diversity of Macroinvertebrates in Two Playa Lakes on the Southern High Plains, Texas, U.S.A." *Southwestern Naturalist* 26:153–58.

Moore, W. G., and A. Burn. 1968. "Lethal Oxygen Thresholds for Certain Temporary Pond Invertebrates and Their Application to Field Situations." *Ecology* 49:349–51.

Neal, J. T., ed. 1975. *Playas and Dried Lakes*. (Benchmark papers in geology; vol. 20). Stroudsburg, Pa.: Dowden, Hutchinson & Ross, Inc.

Noy-Meir, I. 1973. "Desert Ecosystems: Environment and Producers." *Annual Review of Ecological Systems* 4:25–51.

Pennak, R. W. 1978. "Fresh Water Invertebrates of the United States." 2d ed. New York: Ronald Press Co.

Peterson, R. C., and K. W. Cummins. 1974. "Leaf Processing in a Woodland Stream." *Freshwater Biology* 4:343–68.

Proctor, V. W., C. R. Malone, and V. L. DeVlaming. 1967. "Dispersal of Aquatic Organisms: Viability of Dissemules Recovered from the Intestinal Tract of Captive Killdeer." *Ecology* 48:672–76.

Prophet, C. W. 1963. "Physical-Chemical Characteristics of Habitats and Seasonal Occurrence of Some Anostraca in Oklahoma and Kansas." *Ecology* 44:798–801.

Reeves, C. C., ed. 1972. Playa Lake Symposium. Icasals Publication No. 4.

Rzoska, J. 1961. "Observations on Tropical Rainpools and General Remarks on Temporary Waters." *Hydrobiologia* 17:265–86.

———. 1984. "Temporary and Other Waters." In *Key Environments: Sahara Desert*, ed. J. L. Cloudsley-Thompson, 105–27. New York: Pergamon Press.

Saunders, G. W. 1976. "Decomposition in Freshwater." In *The Role of Terrestrial and Aquatic*

Organisms in Decomposition Processes, ed. J. M. Anderson and A. Macfadyen, 341–73. 17th Symposium of the British Ecological Society, Blackwell Scientific Publications.

Sublette, J. E., and M. S. Sublette. 1967. "The Limnology of Playa Lakes on the Llano Estacado, New Mexico and Texas." *Southwestern Naturalist* 12:369–406.

Wetzel, R. G. 1975. *Limnology*. Philadelphia, Pa.: W. B. Sanders Co.

Whitford, W. G. 1973. Jornada Validation Site Report; Desert Biome, U.S. International Biological Program, Research Memorandum 73–4, Vol. 2.

———. 1975. Jornada Validation Site Report; Desert Biome, U.S. International Biological Program, Research Memorandum 75–4, Vol. 2.

Whitford, W. G., and H. Becker. 1972. "Comparative Ecology of Montane and Desert Represos (cattle-watering impoundments) in Southern New Mexico." Final Report to New Mexico Water Resources Research Institute (2144).

Whitford, W. G., and J. A. Ludwig. 1971. Jornada Validation Site Report; Desert Biome, U.S. International Biological Program.

Whitford, W. G., and K. H. Meltzer. 1976. "Changes in O Consumption, Body Water and Lipids in Burrowed Desert Juvenile Anurans." *Herpetologica* 32:23–25.

Whitford, W. G., D. W. Freckman, L. W. Parker, D. Schaefer, P. F. Santos, and Y. Steinberger. 1983. "The Contributions of Soil Fauna to Nutrient Cycles in Desert Systems." In *New Trends in Soil Biology,* ed. P. Lebrun, H. M. Andre, A. DeMedts, C. Gregoir-Wibo, and G. Wauthy, 49–59. Proceedings of the 8th International Colloquium of Soil Zoology, Louvain-la-Neuve (Belgium). Dieu-Brichard, Ottignies-Louvain-la-Neuve, Belgium.

Woodward, B. D. 1983. "Predator-Prey Interactions and Breeding-Pond Use by Temporary Pond Species in a Desert Anuran Community." *Ecology* 64:1549–55.

7

Playa Lakes: Natural Impoundments and Waterfowl Management of the Southern High Plains of Texas

Eric G. Bolen and Paul N. Gray

SUMMARY

In the Southern High Plains of Texas, there are 18,000 playa lakes that provide habitat for wildlife. Man's activities in this region can be both beneficial and detrimental to the playa environments. The irrigation runoff which collects in the clay-lined basins provides a water supply, while the waste corn is the mainstay of the diet of the waterfowl and migratory birds. However, excavation of playas by some farmers to create steep-sided pits (to retain more water) has diminished the biological productivity and ecological diversity of the littoral zones. A study of seven pitted playas (four terraced and three control) designed for water storage found the following: 1) there was no difference between the experimental and control playas in the abundance of invertebrates, although more were available for use by waterfowl in the terraced playas; 2) vegetative production and seed accumulation were greater on the terraced playas; and 3) although the experimental and control playas were used equally by waterfowl in summer and fall, in winter there were more species of birds and greater use of the terraced playas.

INTRODUCTION

In Texas, some 18,000 playa basins dot the landform known as the Southern High Plains (Fig. 7.1). This region, also known as the Llano Estacado or "Staked Plain" since the days of Spanish explorations, is a semi-arid

Figure 7.1. Surface runoff on the Southern High Plains of Texas collects in clay-lined basins known as playa lakes. The playas form "islands" of wildlife habitat in a "sea" of intensive agriculture. Photo courtesy of High Plains Underground Water Conservation District No. 1, Lubbock, Texas.

tableland essentially devoid of natural surface water save for the ephemeral playas. Rivers once traversed the Southern High Plains but geological events following the Pliocene hydrologically isolated the region; surface water once flowing across the 82,000 km^2 area now is diverted eastward by the Canadian River and southward by the Pecos River. Nonetheless, the Ogallala Aquifer remains as a geological legacy from these earlier times and that formation interacts profoundly with the regional economy, playas, and wildlife (Bolen, Simpson, and Stormer 1979; Guthery 1981; Stormer, Bolen, and Simpson 1981; Bolen 1982; Bolen and Guthery 1982; Guthery, Pates, and Stormer 1982).

The playas in Texas are uniquely adapted for impounding surface drainage because of their clay-lined basins. Surrounding soils generally are well-drained sandy loams, whereas the playas developed with nearly impermeable Randall clays. Some tales from yesteryear attribute the depressions to the wallowing of bison (*Bison bison*), but geological studies confirm that most playas originated from the repeated scouring of strong winds sweeping across the prairie landscape (Reeves 1966). With human settlement late in the nineteenth century, ranching and farming shaped out of the fertile soils an agricultural economy.

Today the Southern High Plains is one of the most intensively cultivated regions in the Western Hemisphere, flourishing since World War II from irrigation water pumped from the Ogallala Aquifer. More than 70,000 wells tapped the Ogallala in 1977 (New 1979). However, the aquifer is a mined resource; pumping exceeds the pace of recharge and most of the meager precipitation (about 46 cm annually) trapped in the playas evaporates without entering the aquifer. As a result, estimates suggest that the saturated thickness of the Ogallala in our study area (Castro County) will diminish from its 1974 thickness of 33.5 m to less than 6 m by the year 2020 (Wyatt, Bell, and Morrison 1976). Those projections, in turn, reflect a reduction of 12.8 to 3.3 million acre-feet of stored water in the Ogallala.

PLAYAS AND IRRIGATION

Irrigation has set the stage for several interactions with playa lakes and wildlife. First, irrigation runoff collects in many playa basins, offering water supplies not otherwise available to wildlife. Estimates made by the High Plains Underground Water Conservation District No. 1 (1977) indicate that nearly 20% of the water pumped for irrigation probably becomes runoff leaving the fields. The importance of that runoff to playa ecology, of course, is magnified in years when precipitation is limited. In any case, the influence of tailwater recovery often is significant; interspersion of plant communities may be increased 14-fold in playas receiving irrigation runoff, with the presence of cattail (*Typha* spp.), arrowhead (*Sagittaria* sp.), and pondweed (*Potamogeton* spp.) apparently related directly to tailwater recovery (Guthery, Pates, and Stormer 1982).

Second, farmers have modified the pristine configurations of many playas by excavating steep-sided pits in these small watersheds (Fig. 7.2). According to a recent survey, the number of modified playas increased from 150 to 10,800 in the 15-year period between 1965 and 1980 (Guthery et al. 1981). Guthery and Bryant (1982) determined from an aerial survey that 69% of the playas larger than 4 ha were modified for water storage. With modification, the collected runoff can be recycled from the pits for additional irrigation without the heavy financial burden of pumping "new" water from deep wells. The pits also store runoff in surface-area-to-volume ratios minimizing evaporational losses. Obviously, the ecological impacts vary with the size of each excavation and the size of the lake basins in which these modifications are made. However, the biological productivity and ecological diversity of littoral zones in many playas likely are diminished

Figure 7.2. Pits excavated in playa basins store run-off water for agricultural purposes but, as shown here, drastically reduce littoral zones.

when water is shunted into storage pits instead of inundating the shallow basins. For example, unmodified playas contained significantly larger populations of insects and nearly twice as many families of insects than lakes modified with pits (Rhodes and Garcia 1981; Merickel and Wangberg 1981). Such relationships also were inferred by differences in the ecology of waterfowl feeding on modified and unmodified playas (Rollo and Bolen 1969). The richness and productivity of littoral zones nonetheless remain a cornerstone in the ecology of aquatic environments; virtually all trophic levels in aquatic systems function directly or indirectly from the basic production of littoral zones. Hence, Fredrickson and Taylor (1982) recently proposed a water level management plan for moist-soil vegetation as a resource for birds and other wildlife, but we are only now beginning to develop ecological criteria and a similar proposal for managing playa wetlands.

Third, land use patterns undoubtedly will change as water from the Ogallala diminishes. Pumping costs also continue to escalate, further impeding continued mining of what once was an inexpensive and abundant resource. Two options seem obvious as the dynamics of the water regime come into play. One is to return to dryland farming. The Texas Department of Water Resources estimated that the acreage devoted to irrigated crops (especially corn and wheat) will diminish by 99% in the interval between

1977 and 2020. The other option is to return to a rangeland economy, perhaps not unlike an earlier era when cattle barons grazed their herds on a vast shortgrass prairie. In either scenario, however, the playas will no longer receive irrigation tailwater in amounts known today. Nor will those playas modified for water storage quickly regain their original carrying capacity for wildlife.

WILDLIFE RESOURCES

The playas stand as "islands" of wildlife habitat in a "sea" of intensive agriculture. Waterfowl are foremost among the wildlife populations dependent on playa environments although ring-necked pheasants (*Phasianus colchicus*) represent another important resource. Pheasant numbers average more than 130 birds per playa where corn is the dominant crop (Guthery and Whiteside 1984), and 30–40% of the chicks are produced on playas even though these habitats occupy no more than 4% of the area (Bolen and Guthery 1982).

The Southern High Plains is traversed by the Central Flyway, one of four major pathways for migratory birds in North America (Fig. 7.3). In years when surface water fills the playas, one to two million waterfowl spend five to seven months overwintering in the region. Concentrations of ducks wintering on playa lakes often reach enormous densities (Fig. 7.4). Others continue their journeys, using the playas as resting stops en route to other destinations.

Mallards (*Anas platyrhynchos*), northern pintails (*A. acuta*), green-winged teal (*A. crecca*) and American wigeon (*A. americana*) are foremost in abundance among waterfowl wintering on the Southern High Plains, but sizable numbers (ca. 100,000) of Canada geese (*Branta canadensis*) also overwinter in the region.

Waste corn is the mainstay of the waterfowl diet throughout the winter on the Southern High Plains, again indicating the strong interaction between the current agricultural regime and wildlife. Corn comprised more than 93%, by volume, of the diet for the four major species based on an autumn and winter sample of 647 birds (Moore 1980). One can only speculate how waterfowl wintering on the Southern High Plains subsisted prior to the availability of corn, but to the extent that playas alone once provided the necessary foodstuffs, then it seems unlikely that the lakes can continue for much longer to meet the needs of the current waterfowl population without intensive management. Reduced irrigation means less corn production and reduced tailwater accumulation that, together with modifications

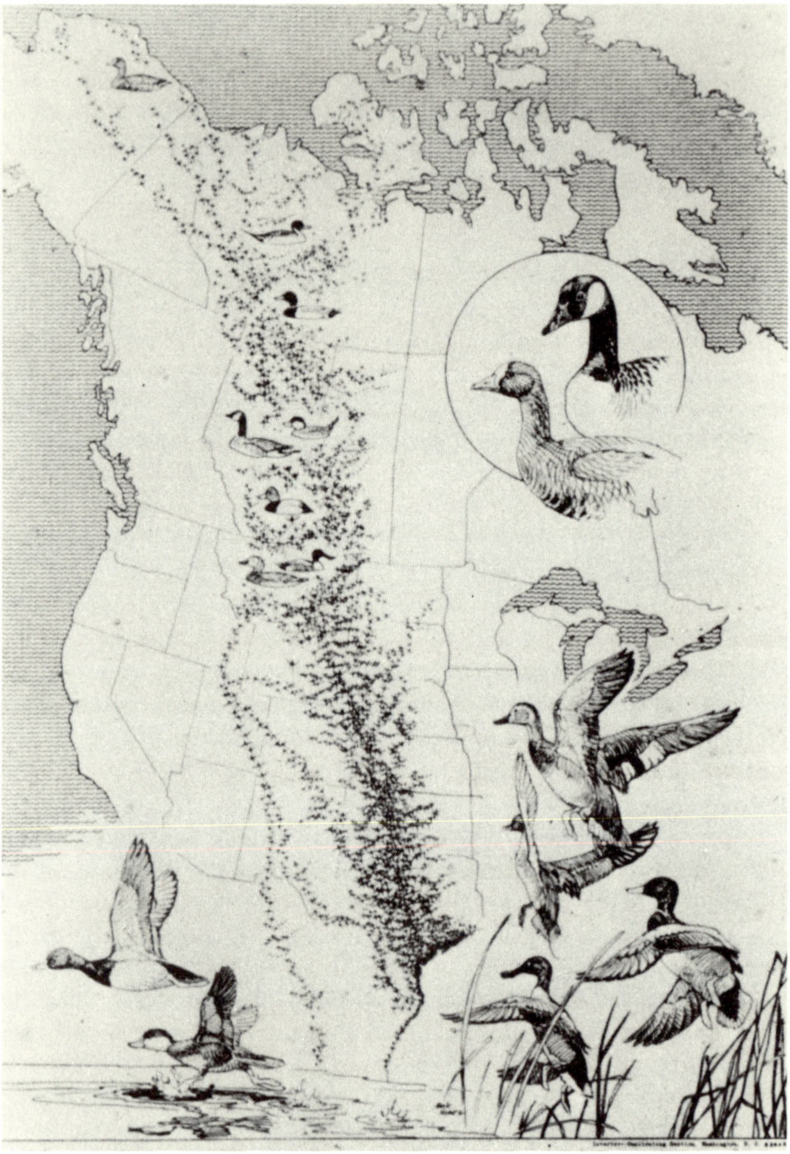

Figure 7.3. The Central Flyway is one of four major routes for migratory birds in North America. Note that the flyway passes over the Southern High Plains of Texas where some one to two million waterfowl overwinter on playa lakes. Photo courtesy of U.S Fish and Wildlife Service.

Figure 7.4. Large numbers of waterfowl overwinter on playa lakes.

of the basins, seem certain to alter the regime now supporting large numbers of wintering waterfowl.

For the present, detailed studies of the availability and abundance of waste corn in relation to post-harvest treatments of croplands offer a means of managing the winter diet of field-feeding waterfowl (Baldassarre et al. 1983). Where disking reduces the abundance of waste corn by 77%, burning the ground litter leaves virtually all waste corn at maximum abundance and availability. Field-feeding waterfowl make two daily flights for waste corn, often "keying" onto fields no longer covered with surface litter (Baldassarre and Bolen 1984). In general, mallards remain in good body condition in winter because of their use of waste corn (Whyte 1983). However, despite the high energy content of corn, such a monotypic diet is not balanced nutritionally. The amino acids lysine, tryptophan, and methionine are lacking in corn yet remain essential in the diets of higher vertebrates (Schaible 1970; Sturkie 1976). Hence, protein-rich foods presumably produced in the littoral zones of playas also appear crucial to the well-being of wintering waterfowl. Waterfowl of both sexes experiencing various molts require protein as do hens entering breeding condition; Heitmeyer (1983) determined that mallard hens begin consuming invertebrates in early spring while still on their wintering grounds. Quinlan and Baldassarre (1984) also noted that green-winged teal consumed natural foods (i.e., invertebrates and seeds of native plants) throughout their wintering period on the Southern High Plains of Texas. Such foods were regarded as an important supplement offsetting the nutritional deficiencies associated with a diet otherwise dominated by waste grain. Thus, the continued availability of proteinaceous foods from those playas modified for water storage seems a crucial goal for management.

With this background in mind, some preliminary results stemming from an experiment in habitat management are of interest. The setting involves a management strategy proposing to enhance habitat values for waterbirds in playas where (1) pits may be excavated in the future or where (2) pits already may have reduced or eliminated functional littoral zones. In the latter instance, future trends are anticipated, namely reduced irrigation and lessened volumes of tailwater. Hence, the experiments may produce a means of reclaiming playa environments at a time in the future when precipitation again may provide the only source of surface water on the Southern High Plains.

ARTIFICIAL LITTORAL ZONES

Seven playas located in Castro County, Texas, were selected as experimental units. Of these, four were excavated with a pit designed for water storage except that a series of terraces were placed on one side. The other three playas already were pitted but were reexcavated on one side in a similar fashion. In general, the terraces were about 70 m long and 8 m wide; each was arranged in a staircase, with about 45 cm between each step (Fig. 7.5). The three previously pitted playas served as controls.

Sampling included measurements of (1) use by ducks and other waterbirds as determined by twice weekly censuses, (2) benthic invertebrates and residual seeds ("seed bank") from 7.5 cm cores, (3) swimming and plant-clinging invertebrates with net sweeps, and (4) vegetation development by height, percentage groundcover, composition, and biomass in October 1983.

Each site was designated for identification as T (for terraced) or as C (for control) and numbered consecutively.

RESULTS

The studies are yet incomplete, and the results are preliminary. Hence, the statistical analyses are unsophisticated at present and compare terraced pits with controls using Student's "t" test (Steel and Torrie 1980). Nonetheless, some trends have emerged that, barring reversal as additional work is completed, suggest positive responses.

Figure 7.5. Terraces constructed on one side of a pitted playa basin offer artificial littoral zones for wildlife under a regime of fluctuating water levels. Shown here is experimental unit T1. Note dense strip of *Polygonum* on terrace at far right; another terrace is under 10 cm of water at far left. Temporarily barren terraces offer a feeding habitat for shorebirds but soon will support semi-aquatic vegetation.

Invertebrates

No differences ($P > 0.05$) have occurred to date in the abundance of invertebrates between terraced and control pits (Table 7.1). *Daphnia* blooms developed in pits T1, T2, and C4. Pit T2, because of its alternating wet-dry cycle, also developed blooms of fairy shrimp (*Anostraca* sp.).

Overall, chironomid larvae have remained the most common invertebrates in the samples. They require inundated habitat and thus are abundant on the more regularly flooded parts of terraced and control pits. However, the ecological difference between abundance and availability should be emphasized in reporting these and other data. Specifically, chironomid larvae, while no less abundant in either type of pit, occur in deeper water (1–3 m) in the control pits and therefore remain largely beyond the feeding range of mallards and other dabbling ducks. Conversely, chironomid larvae and other foods occurring on the terraces are covered by water within the foraging depth of most dabbling ducks. Ideal foraging conditions for puddle ducks occur in water less than 30 cm in depth (Fredrickson and Taylor 1982). In the experimental pits, designed with increments of 45 cm

Table 7.1. Mean numbers of invertebrates in terraced and control pits based on samples obtained in October 1983

	Mean No. Per Core[a]		Mean No. Per Net Sweep[a]	
Pit No.	Terraced	Control	Terraced	Control
1	1.6	—[b]	15.0	—[b]
2	0.3	5.0	61.3	0.6
3	0.9	0.9	0.0	0.1
4	0.8	0.6	0.4	29.0

[a]No significant differences ($P > 0.05$) between means for terraced and control pits for either core samples (df = 91) or net sweeps (df = 86).
[b]Not sampled.

between terraces, water depths at any time (i.e., increases during collection of runoff or decreases during pumping) offer puddle ducks more opportunities for feeding on proteinaceous foods. Moreover, in terraced pits, about 20% of the area of surface water was less than 30 cm deep, whereas in a control pit about 10% of the water was of a similar depth. Hence, terraced pits offered twice the feeding area for waterfowl.

Plant Production

Vegetation responded dramatically to growing conditions on the terraces. Plant production was greater ($P < 0.05$) on terraced pits than on control pits (Table 7.2). Production was greatest on those terraces that experienced a regime of alternate flooding and drying. On average, about two terraces

Table 7.2. Mean plant production (g/m^2) on terraced and control pits based on samples obtained in October 1983[a]

Pit No.	Terraced[a]	Control
1	76	—[b]
2	1,052	28
3	320	20
4	—[b]	20

[a]Means for terraced pits are significantly different ($P < 0.05$, df = 56) from means for controls.
[b]Not sampled.

per pit produced excellent stands of vegetation whereas other terraces in the "staircase" often lacked similar development. Thus, given the regime of alternating water levels resulting from agricultural activities, the presence of several terraces (i.e., a full "staircase") assured good plant production under virtually all conditions. Conversely, plant production on the control pits was restricted to a narrow zone along the shoreline.

Vegetation developing on either the terraces or on the narrow zones on the periphery of control pits largely consisted of species considered as valuable waterfowl foods. About 92% of the vegetation on T1 was one or more species of *Polygonum, Echinochloa,* or *Eleocharis.* On T2, *Polygonum, Echinochloa, Kochia,* or *Ambrosia* covered 91% of the terraces. Pit T3 was wetter than the other terraced pits, and *Chara* covered about 50% of the site, with *Cyperus* and *Salix* covering an additional 40% of the area.

On the control pits, *Echinochloa* comprised about 45% of the cover; four other plants each accounted for about 10% of the area covered by vegetation. Seed banks did not differ significantly ($P > 0.05$) between terraced and control pits although, on average, somewhat larger seed accumulations occurred on terraced pits after only one growing season (Table 7.3). This suggests that sampling in subsequent seasons may reveal dramatic differences since the control pits are older, yet apparently have smaller seed banks after several years of potential seed accumulation than shown on the one-year-old terraces.

Responses of Waterbirds

Birds also responded dramatically to the artificial littoral zones on the terraced pits. Significantly more ($P = 0.029$) species of birds were observed on terraced pits than on control pits (Fig. 7.6). These included shorebirds and herons as well as waterfowl.

Table 7.3. Mean weight (g) of seeds per core on terraced and control pits based on samples obtained in October 1983

Pit No.	Terraced[a]	Control
1	0.09	—[b]
2	0.13	0.08
3	0.01	0.01
4	—[b]	0.01

[a]No significant differences ($P > 0.05$, df = 64) between means for terraced and control pits.
[b]Not sampled.

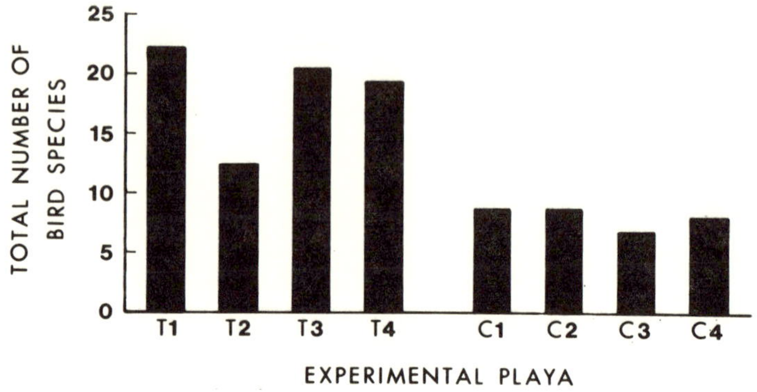

Figure 7.6. Number of bird species censused in 1983 on terraced pits (T1–T4) was significantly greater (P = 0.029) than on control pits (C1–C4) (Wilcoxon 2–Sample test).

Terraced pits attracted more shorebirds ($P < 0.05$) than occurred on control pits (Table 7.4). Further, 13 species of shorebirds visited the terraced pits whereas 3 species were censused on the control pits. Presumably, mudflats available on the terraces created attractive feeding habitat for shorebirds whereas little habitat of that kind existed along the shorelines of control pits.

Significantly more ($P < 0.05$) waterfowl used terraced pits than control pits in late winter (Table 7.5). Conversely, waterfowl used all pits in about

Table 7.4. Mean number of shorebirds per census on terraced and control pits during spring and fall 1983

	Terraced				Control			
Season	1	2	3	4	1	2	3	4
Spring[a]	6.0	3.7	0.5	—[b]	0.1	1.4	0.0	—[b]
Fall[a]	5.1	2.1	1.3	7.4	0.9	0.1	0.4	1.2

[a]Means are significantly different ($P < 0.05$) between terraced and control pits for both spring (df = 49) and fall (df = 121) censuses.
[b]Not sampled.

Table 7.5. Mean number of ducks per census on terraced and control pits during summer, fall, and late winter 1983

	Terraced				Control			
Season	1	2	3	4	1	2	3	4
Summer	2.9	0.0	0.1	0.0	1.7	0.5	0.2	—[b]
Fall	2.5	2.8	10.6	9.2	6.4	5.0	0.0	0.8
Late winter[a]	254.0	0.0	334.0	451.0	—[b]	55.0	0.0	78.0

[a]Means are significantly different ($P < 0.05$, $df = 65$) between terraced and control pits only in late winter, not significant in summer ($df = 150$) or in fall ($df = 147$).
[b]Not sampled.

equal numbers ($P > 0.05$) during summer and autumn months when ducks either were breeding elsewhere in the Central Flyway or gathering in large flocks on playas of greater sizes than the study sites. Since the study area primarily serves as a wintering area for waterfowl, the results to date suggest that terraced pits enhance winter habitat in ways that benefit ducks overwintering on the Southern High Plains of Texas. As with shorebirds, the census revealed more species (14) of waterfowl on terraced pits than occurred on control pits (8).

The diversity and abundance of shorebirds and waterfowl on the terraced pits serve as indicators of habitat quality. Herons offer another example since none was included in the censuses of any pits lacking fish populations. Of the eight pits included in the results to date, five contained fish and only on these were herons observed (although heron tracks were noted on the others). In this case, pits lacking fish did not meet the immediate needs of herons. Hence, whether herons or other birds stay and invest their time and energies for food on a given pit provides a measure of habitat preference; that is, the birds themselves become a "panel of experts" for habitat quantity and quality.

Based on preliminary data, invertebrate populations were about equal on terraced and control pits. However, pits with terraces produced more vegetation and tended to attract more species and numbers of waterbirds. Terraces increased the availability of proteinaceous foods (invertebrates and seeds of native plants) for birds because a littoral zone of nearly twice the size is created in comparison with unterraced (control) pits.

ACKNOWLEDGMENTS

The experimental work is funded by the Texas Parks and Wildlife Department as part of Federal Aid Project W–106–R. In particular, we thank William C. Brownlee, Charles D. Stutzenbaker, Ted L. Clark, and Bobby G. Alexander for their encouragement and support of our endeavors. Additional support was provided by Organized Research for Agriculture, Texas Tech University. We also acknowledge assistance in the field and/or laboratory from Richard J. Whyte, Carol A. Gray, Chris E. Gregory, Michael J. Davin, Duc Chi Nguy, Louis J. Celentano, Sang-don Lee, and Brian L. O'Kelley. The manuscript benefited from reviews by Guy A. Baldassarre and Fred S. Guthery. This is paper T–9–380, College of Agricultural Sciences, Texas Tech University.

REFERENCES

Baldassarre, G. A., and E. G. Bolen. 1984. "Field-Feeding Ecology of Waterfowl Wintering on the Southern High Plains of Texas." *Journal of Wildlife Management* 48(1):63–71.

Baldassarre, G. A., R. J. Whyte, E. E. Quinlan, and E. G. Bolen. 1983. "Dynamics and Quality of Waste Corn Available to Postbreeding Waterfowl in Texas." *Wildlife Society Bulletin* 11(1):25–31.

Bolen, E. G. 1982. "Playa Wetlands of the U.S. Southern High Plains: Their Wildlife Values and Challenges for Management." In *Wetlands Ecology and Management*, ed. B. Gopal, R. E. Turner, R. G. Wetzel, and D. F. Whigham, 9–20. New Delhi, India: National Institute of Ecology and International Scientific Publications.

Bolen, E. G., and F. S. Guthery. 1982. "Playas, Irrigation, and Wildlife in West Texas." *Transactions of the North American Wildlife and Natural Resources Conference* 47:528–41.

Bolen, E. G., C. D. Simpson, and F. A. Stormer. 1979. "Playa Lakes: Threatened Wetlands of the Southern Great Plains." In *Riparian and Wetland Habitats of the Great Plains*, Publication 91, Great Plains Agricultural Council. Fort Collins, Colo.: USDA Forest Service.

Fredrickson, L. H., and T. S. Taylor. 1982. "Management of Seasonally Flooded Impoundments for Wildlife." *USDI Fish and Wildlife Service Resources* Publication 148. Washington, D.C.

Guthery, F. S. 1981. "Playa Basins and Resident Wildlife in the Texas Panhandle." In *Proceedings of the Playa Lakes Symposium*, ed. J. S. Barclay and W. V. White, 47–51. Report FWS/OBS–81/07. Washington, D.C.: USDI, Fish and Wildlife Service.

Guthery, F. S., and F. C. Bryant. 1982. "Status of Playas in the Southern Great Plains." *Wildlife Society Bulletin* 10(4):309–17.

Guthery, F. S., F. C. Bryant, B. Kramer, A. Stoecker, and M. Dvoracek. 1981. "Playa Assessment Study." Report submitted to Water and Power Resources Service, Amarillo, Tex.

Guthery, F. S., J. M. Pates, and F. A. Stormer. 1982. "Characterization of Playas of the North-Central Llano Estacado in Texas." *Transactions of the North American Wildlife and Natural Resources Conference* 47:516–27.

Guthery, R. S., and R. W. Whiteside. 1984. "Playas Important to Pheasants on the Texas High Plains." *Wildlife Society Bulletin* 12(1):40–43.

Heitmeyer. 1983. Personal Communication.
High Plains Underground Water Conservation District No. 1. 1977. "Guide to Irrigation Tailwater Recovery." Report 77–01. High Plains Underground Water Conservation District No. 1, Lubbock, Tex.
Merickel, F. W., and J. W. Wangberg. 1981. "Species Composition and Diversity of Macroinvertebrates in Two Playa Lakes on the Southern High Plains, Texas." *Southwestern Naturalist* 26(2):153–58.
Moore, R. L. 1980. *Aspects of the Ecology and Hunting Economics of Migratory Waterfowl on the Texas High Plains.* Master's thesis, Texas Tech University, Lubbock.
New, L. 1979. *High Plains Irrigation Survey.* College Station: Texas Agricultural Extension Service.
Quinlan, E. E., and G. A. Baldassarre. 1984. "Activity Budgets of Nonbreeding Green-Winged Teal on Playa Lakes in Texas." *Journal of Wildlife Management* 48: in press.
Reeves, C. C., Jr. 1966. "Pluvial Lake Basins of West Texas." *Journal of Geology* 74(3):269–91.
Rhodes, M. J., and J. D. Garcia. 1981. "Characteristics of Playa Lakes Related to Summer Waterfowl Use." *Southwestern Naturalist* 26(3):231–35.
Rollo, J. D., and E. G. Bolen. 1969. "Ecological Relationships of Blue and Green-Winged Teal on the High Plains of Texas in Early Fall." *Southwestern Naturalist* 14(2):171–88.
Schaible, P. 1970. *Poultry: Feeds and Nutrition.* Westport, Conn.: Avi Publications.
Steel, R. D., and J. H. Torrie. 1980. *Principles and Procedures of Statistics.* New York: McGraw-Hill Book Co.
Stormer, F. A., E. G. Bolen, and C. D. Simpson. 1981. "Management of Playas for Migratory Birds—Information Needs." In *Proceedings of the Playa Lakes Symposium*, ed. J. S. Barclay and W. V. White, 52–61. Report FWS/OBS–81/07. Washington, D.C.: U.S. Fish and Wildlife Service.
Sturkie, P. D., ed. 1976. *Avian Physiology.* New York: Springer-Verlag.
Whyte, R. J. 1983. *Winter Condition of Mallards on the Southern High Plains of Texas.* Ph.D. diss., Texas Tech University, Lubbock.
Wyatt, A. W., A. E. Bell, and S. Morrison. 1976. *Analytical Study of the Ogallala Aquifer in Castro County, Texas.* Texas Water Development Board Report 206, Texas Department of Water Resources, Austin, Tex.

8

Water Impoundments in the Arid Environment

Jon E. Rodiek

SUMMARY

Because land use activities in arid environments require additional water and energy, planning decisions must be based on dependable knowledge of the costs and benefits of providing these resources. The land use planner today is challenged to provide immediate solutions to problems and to contribute scientific information that will benefit society. Increased demands of (and the subsequent costs of) water in arid regions have forced readjustment of societal values regarding water allocations for the five most common water use classes: livestock use, irrigation storage, fish production, wildlife use, and multiple-purpose use. Project planning for impoundments that capture, store, and deliver water requires initially an assessment of the needs, of the resources that already exist, and of the opportunities for improvement. Sometimes a preliminary study that identifies additional requisite investigations is needed for impoundment design and construction. Predictions and decisions should be based on collaborative information from professionals and made with the knowledge that water conservation now has a significant influence on water use plans in the arid West. Three projects in Arizona were designed to use previously unused water and to redistribute it to the environment. The projects used water from urban and rangeland runoff as well as sewage effluent to control erosion, provide recreational opportunities, benefit wildlife, control channel erosion, and dispose of effluent in locations other than streams. An increased awareness of water as a precious resource in arid areas will stimulate additional projects such as these in the future.

INTRODUCTION

The resource planner who would seek to achieve sustained land use in the arid environment must recognize some basic realities. The arid environment (this discussion of arid environments is confined to the hot, dry desert regions of the world) is one of low biological productivity despite being immersed in a high solar energy field. The natural biological system is able to use the energy and water resources found there in unique fashion. We do not possess a technological equivalent within our resource planning system; for this reason we must subsidize our land use activities with added water and energy supplies. When we import and concentrate water resources, we are not changing this reality. We are simply relying on our own socio-economic resources in order to realize gains in land use production in the arid environment, although the cost is high. The most common error made in evaluations of such arid-land projects is overestimating or miscalculating the actual level of return. Therefore, it is imperative to base planning decisions on dependable knowledge in order to realistically assess the cost and benefits of the planned improvements.

A water impoundment project located in the arid environment has a very definite life expectancy. The environment quickly reclaims any water impoundment that has been developed with no allowance for the low amounts of water available and the high water losses that can occur. It is essential for the planner to carry out a sound analysis of the baseline conditions of the area in question.

There is a question of appropriateness in arid-land water impoundment projects. Understanding how successful or unsuccessful past plans for such projects have been is a primary improvement that needs to be made in our current analyses. Poorly conceived water impoundment projects can convert enthusiasm into skepticism. A major criticism is that we do not have a good model of how the man-manipulated resource process in the arid environment really works. It behooves resource planners to concentrate on the experimental versions of more grandiose projects in the past as they attempt to determine workable solutions for the project at hand. The planners must recognize there is no simple path to the efficient use of available resources in the arid environment. Much more basic research needs to be done to realize this goal. The concept of developing small water impoundments in arid environments is valid if it is conducted in the spirit of improving the decision-making ability of planners and to supply the basic human needs of our society. In this sense, every project becomes a research experiment that can add to our collective knowledge of the development of the arid environment.

Planning for water impoundments in arid regions requires the coordination of two kinds of goals. First, the project must be clearly defined in terms of its client. By doing so, the planners are able to determine whether or not a reasonable solution can be worked out with the resources of personnel, money, land, water, and materials at their disposal. Also the project must be set into a proper philosophical context so that each project contributes knowledge to the larger problems of resource development in the arid landscape. Such efforts supply society with powerful weapons in our search for solutions to the efficient use of resources.

Water resources, even though they are finite, will continue to be developed in the arid landscape. The purpose of this chapter is to demonstrate through case study analysis the various techniques and procedures used to produce three distinct water impoundments. The discussion is organized under three topic headings: 1) Assessing Present Uses and Impoundment Types, 2) Planning the Water Impoundment, and 3) Case Study Analysis.

ASSESSING PRESENT USES AND IMPOUNDMENT TYPES

A water impoundment project that provides scientific information and yet fails as a practical land use plan cannot be perceived as a valid project by a land planner. Land planners are people interested in practical and intellectual solutions that improve problem-solving methods and people's lives. A key to the initiation of a procedure for a valid solution is a review of existing projects that are similar in character to the proposed project. Since arid-land impoundment projects are now being developed throughout the arid West, the review which follows of selected projects revealing five classes of water use and three types of impoundments should prove useful to planners working in arid areas.

USES FOR ARID-LAND WATER

Classes of Water Use

Five classes were identified (USDA 1982). These uses include water for livestock, irrigation storage, fish production, waterfowl and wildlife requirements, and general multiple-use purposes.

Livestock. Water is as important as forage in the production of livestock. Particular features of water use management for livestock on rangelands and pastures involve critical amounts and the proper distribution of water throughout the livestock's feeding habitat. An impoundment should be planned so that it contributes to the network of water supplies that services livestock within a particular watershed drainage area. Table 8.1 identifies the average daily consumption of water by various livestock species (USDA 1982). These general estimates must be refined by the planner in view of specific livestock needs and distribution within the feeding habitat.

Irrigation Storage. Water supplies developed for irrigation purposes are the most demanding of all water uses discussed here. An impoundment system must consider the water requirements of the plant species to be irrigated, the effective available rainfall anticipated over the course of the growing season, the evaporation potential of the area, seepage losses, and the distribution efficiency of the irrigation methods. While new technologies in irrigation methods have greatly increased the effective available water supplies used for crop production, reducing losses is as important as increasing available supplies in the overall scheme of resource uses in the arid region. Table 8.2 identifies some general needs for different crops within the 10–20 in. range of precipitation.

Small reservoir impoundments may be built for special needs of crop production such as fertilizing and insecticide and herbicide spraying. The amounts needed for spraying are relatively low but critical when needed. Field crops usually require between 100–150 gal per acre. Orchards may require as much as 1,000 gal or more per acre for each spraying application. Consequently impoundments should be located so as to facilitate delivery to application vehicles. Such impoundments could easily serve as multiple-purpose reservoirs.

One should not rely on small impoundment reservoirs to function as

Table 8.1. Daily consumption of water by livestock

Livestock Type	Gallons Per Head Per Day
Beef cattle and horses	12–15
Dairy cows	15, 35 (dairy and barn needs)
Hogs	4
Sheep	2

Source: USDA (1982) Handbook No. 590.

Table 8.2. Capacity guide for small irrigation reservoirs

Climate	Annual Rainfall	Acre Feet Per Acre Irrigated		
		Vegetable crops	Field crops	Perennials
Semi-arid	10–20″ annually	3.0	4.0	6.0
to				
Arid	under 10″ annually	small reservoirs not reliable		

Source: U.S. Bureau of Reclamation (1973), *Design of Small Dams.*

primary water supply sources for large irrigated acreage projects. Any irrigation project of substantial size (> 40 acres) in arid zones should have a separate water budget plan developed in response to local surface water and groundwater supplies.

Fish Production Uses. Small impoundments can also provide a supplemental source of food protein via fish production. These impoundments can provide 100–300 lbs of fish annually per acre of such water (USDA 1982). Optimal impoundment sizes range from a quarter-acre to several acres in size. These impoundments can easily serve multiple-use purposes given the proper design criteria and a master plan approach.

Waterfowl-Wildlife Uses. Waterfowl and wildlife living in the managed arid environment will become increasingly dependent upon man-made water impoundments for their primary source of water. Their needs and our desire to maintain a viable population of diverse native wildlife and migratory waterfowl species both depend upon our continued support for their future habitat requirements. Water will remain a primary need for many of those species. Currently our interests for these species is predominantly recreational and aesthetic. Soon we may find more tangible economic and agricultural values to be derived from some of these native and migratory species.

Multiple-Purpose Uses. Water impoundments created strictly for single-use purposes in arid regions are many times considered to be a misappropriation of the resource. A single use of the water can be justified if other, related purposes are also included in the cost-benefit analysis. Genuine multiple-use impoundment plans must include an accurate accounting of all water supply needs. These volumes must be taken into account when the total storage volume capacity of the impoundment is determined. In addition, it is important to consider the quality of water needed for each use, compatibility for safety and containment of the different waters, and the ultimate coordination that will manage such combination activities. While economies of scale will largely dictate the scope of any multiple-use water impoundment project, it may be nonetheless important to explore potential combinations that might be useful in future arid-land use scenarios.

Types of Impoundment

Any plan designed to develop water supplies in the arid environment must define the concept of the water capture, storage, and delivery. The impoundment can encompass all three elements or simply the storage component of a more sophisticated water supply system. Water impoundments have been separated into classes based upon their functional role in a particular capture, storage, and delivery scheme (United Nations ECAFE 1962; Sherard, J. L., et al. 1963; USDA 1982; U.S. Bureau of Reclamation 1973; USDA 1973a; Frasier , G. W., and L. E. Meyers 1983; Lauritzen, C. W. 1960; USDA 1973b; Chakroff, M. 1976).

This diverse collection of impoundment types can be separated into three classes based upon their particular function in the water supply system (Table 8.3). Class I serves primarily to store water. Class II serves to capture and store water. Class III serves to capture, store, and facilitate delivery of water.

PLANNING THE WATER IMPOUNDMENT

An arid-land water impoundment project should focus on those land use plans where real gains can be achieved either through increased profits or, more importantly, reduced operational costs. In the past, technology and science have helped us maintain hard-earned gains with respect to arid-land water impoundment planning. The potential for future development may hinge on our ability to link water use projects to industry, urban, and agricultural projects already in place. A first step for this kind of planning begins with an assessment of what already exists both in the natural and man-made landscapes. From this perspective we can determine what opportunities are left open in terms of available water supplies.

Once the project has been clearly defined in terms of the client's needs and resources (technical, human, financial) the planner can then focus attention on the physical resource questions. These questions invariably include 1) the water use–water storage patterns in the watershed, 2) water storage

Table 8.3. Impoundment types

Class	Function			Potenial Uses
	Capture	*Storage*	*Delivery*[a]	
Class I: Simple storage	By gravity flow, surface runoff sources	Simple storage either man-made or natural impoundment	Passive	Livestock, irrigation, fish production, multiple use (combination of any uses).
Class II: Complex capture and storage	By gravity flow and/or pumping, either surface or ground water sources	Man-made and/or natural impoundments	Passive and some remote	Irrigation, multiple uses, special combination uses; no restrictions between uses other than physical separation of users.
Class III: Complex capture, storage, and delivery	Pumping, catchment, and gravity flow of surface and/or ground-water sources	Man-made storage	Remote	Specialized uses requiring water delivery and treatment restrictions.

[a]Passive means use at storage site; remote means use away from storage site.

capacity and capability of a given site, and 3) the prediction of realistic water supplies available for the chosen storage operation.

A preliminary investigation should be carried out to help the planner determine the context in which the actual impoundment will be placed and to identify more specific studies necessary to write the specifications for the design and construction of a selected impoundment type (Table 8.4).

Water use and water storage patterns can be analyzed by surveying the currently active water management and planning agencies in order to identify existing and proposed projects that may affect a proposed site. A survey of the alternative sites from remotely sensed data, aerial surveys, and ground surveys can help substantiate the "as is" condition for the proposed sites.

The actual capability of a site to capture and store water at any given point is a function of the runoff realized at a specific point and the impoundment storage capability. The planner surveys land slope, soil utilization rates, vegetal cover, and other surface covers to determine the capability of the drainage area to provide runoff for an impoundment. Storm characteristics must be documented in terms of amounts, intensities, and duration in order to be able to develop construction specifications.

Of course no one method of analysis can precisely and accurately predict water volumes available at a given point within a watershed; the prediction is based on an extrapolation of what we know and what we believe will take place. Therefore, it is imperative that any predictions not be spurious ones but rather predictions based on a collaborative effort among a consortium of professionals.

The basic purpose of this discussion is to demonstrate and evaluate the design concepts used for land and water resources to link the use of water to the overall purpose of sound, multiple-resource use in the arid environment.

CASE STUDY ANALYSIS

Before studying the three case studies, one should consider two points critical to the assessment of arid-land water impoundments. First, until very recently, our water planning and management institutions have not made the conservation of water either a public or collective social responsibility. As water has become more scarce, water rental and water rights markets have emerged in many political jurisdictions. It is safe to say that water conservation strategies now have a significant influence on all water use plans in the arid West.

Public institutions mandated to regulate the use of water are now encour-

Table 8.4. Preliminary investigations

Investigation	Purpose	Procedure	Data Produced	Value
Water use—Water storage patterns	Identify existing and proposed water impoundment projects in study area	Air photo survey Air survey Ground survey Agency survey	Quantitative documentation of numbers, location's size, types of impoundment using water in study area	Feasibility of proposed water impoundment in a regional context
Water storage capacity	Identify actual water impoundment supply needs	Precipitation survey Storm event survey Land use cover survey Slope analysis Evapotranspiration rates survey	Minimum impoundment depth Storm runoff volume Evaporation-seepage losses Volume needed	Prediction of water storage size, structure, specifications
Water availability	Identify runoff capture potential	Hydrological study of study site	Peak discharge Rate prediction	Prediction of water volumes potentially available

aging more efficient use of water and policies that select for the socially desirable allocation of water. In other words, we have rapidly evolved from perceiving the resource of water as a non-valued commodity to a highly valued one.

Common resources such as water still lie outside the operation of commercial markets. It is difficult therefore to assess the true value of arid-land water impoundment projects based on some traditional economic or social systems model. This fact does not suggest the projects are worthless—quite the contrary. These projects have advanced our scientific knowledge and expanded our understanding of the man-impacted arid environment. These projects additionally represent a substantial investment in arid-land planning based on the promise that they are a form of social insurance against long-term water supply problems. However, until we are able to incorporate the net worth of these projects into a traditional systems model of assessment, the future for more projects of this type will remain uncertain.

The three projects discussed below had two common objectives that reflected this new attitude towards water. Each project was designed to make use of the existing water supply in a way that did not curtail the supplies of primary users. In other words, these projects were able to use the available supply of water in the environment that had previously been unused. Furthermore, these projects allocated the use of water to purposes that would benefit the environment in a socially acceptable manner. Each project devised a system to capture unused water supplies and convert them to forms that made people aware of the positive effects water can have in an arid environment. These projects were, in other words, not used simply for exploitative purposes. They were designed to allow for the redistribution of the water supply to the environment rather than for direct consumptive purposes. In this sense, all three projects demonstrated the principle of water conservation as it applies to the redistribution of use rates that favor not only the present generation but future generations as well.

Project Summary Descriptions

The three projects included the Santa Cruz River Park in Tucson, Arizona; the Bog Hole Water Impoundment near Sonoita, Arizona; and the Pintail Lake project near Show Low, Arizona. Each project has distinct characteristics in terms of its elevation, precipitation, evaporation losses, and growing season. The precipitation-evaporation comparisons reveal that each project is classified as one located in an arid or semi-arid climate (Table 8.5).

Table 8.5. Environmental profile

Project	Elevation (ft/m above sea level)	Precipitation (in./cm/yr)	Evaporation (in./cm/yr open water)	Growing Season (32°F days)
Santa Cruz River Park	2500 ft/762 m	11 in./27.9 cm	70 in./178 cm	220 days
Bog Hole Impoundment	5000 ft/1500 m	20 in./50.8 cm	55–63 in./ 140–160 cm	172 days
Pintail Lake	6500 ft/1981 m	18 in./45.7 cm	55 in./140 cm	110 days

Santa Cruz River Park. The Santa Cruz River Park project was the largest and most ambitious project in terms of land area and investments (Fig. 8.1). Water supplies were to be realized from urban runoff and in-city desert runoff sources. The project's primary purpose was to control urban erosion and develop recreational opportunities on land which was previously abandoned and neglected. The water supplies would be used in a variety of ways to develop recreation and wildlife habitat sites along the riverway. The primary developer was the city of Tucson. The citizens of Tucson were targeted to be the primary beneficiaries. Wildlife native to the area was also perceived to be a potential beneficiary once vegetative cover was reestablished (Table 8.6).

Bog Hole Water Impoundment. The Bog Hole Water Impoundment was the most remote project (Fig. 8.2). Water supplies were to be realized strictly from rangeland runoff generated in the watershed. The lands were part of the Coronado National Forest grazing allotment in Santa Cruz County. This project's primary purpose was to provide waterfowl habitat for native duck species. The project was also designed to control the widespread channel erosion on these rangelands. The impoundment is approximately 9.2 acres (3.7 ha) within a 198-acre (80 ha) enclosure. In addition to supporting aquatic and wetland habitats, the water supplies provide local livestock and rangeland wildlife with reliable water year-round. The Coronado National Forest and the Arizona Game and Fish Department cosponsored the project.

Pintail Lake. The Pintail Lake project is located at the highest elevation. Its elevation, 6,500 feet (1,981 m), and precipitation, 18 in. per year (45.7 cm) are accompanied by harsh cold winters and drying winds (Fig. 8.3). Soils are heavy and difficult to cultivate for even the most modest rangeland activities. Water supplies were to be derived from sewage effluent pumped by underground pipe to the site from the city of Show Low, Arizona. The project's purpose was to provide an alternative system of disposing of the city's sewage effluent other than in streams. The city's growth demanded an expansion of existing facilities. The city wanted to develop a more cost-effective way to dispose of the effluent while keeping intermittent streams free from waste disposal dumping. Water uses were restricted to a series of three settling ponds, 57 total acres (23 ha) and 15 nesting islands for waterfowl. A total of 500,000 gal per day (1,890,000 l) of secondary treated effluent could be pumped here. Evaporation rates were calculated at one acre-foot per day. Therefore, runoff was not anticipated. The citizens of Show Low were targeted as the primary beneficiaries because of the resulting lower tax rates on municipal services and an added recreational site. Waterfowl and rangeland wildlife benefited significantly as well. The EPA, the Lakeside Ranger District, Apache-Sitgreaves National Forest, and the city of Show Low cosponsored this project.

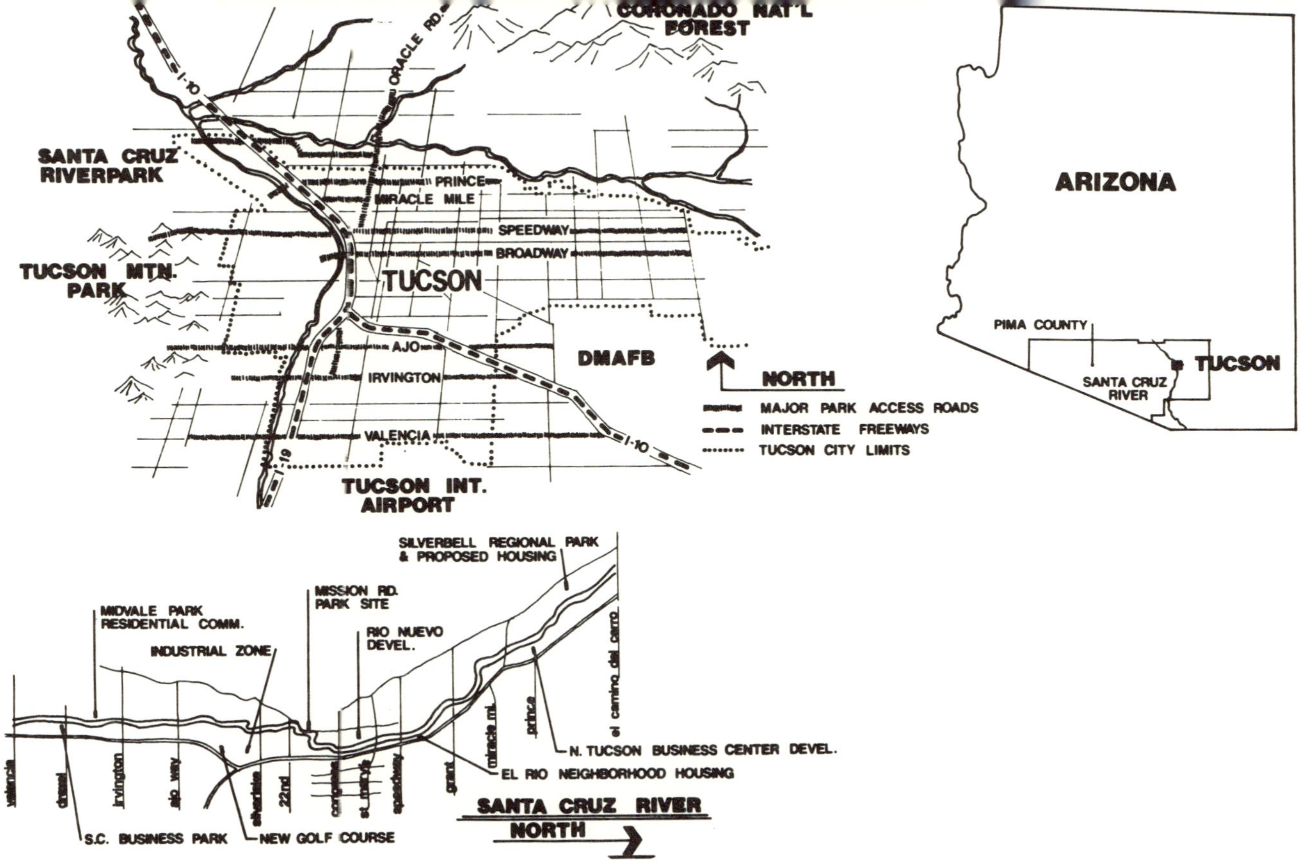

Figure 8.1. Santa Cruz River Park.

Table 8.6. Project Profile

Project	Water Source	Project Class	Water Uses	Project Purpose	Cooperators	Beneficiaries
Santa Cruz River Park	Urban runoff	Class III	Recreation	Erosion control	City of Tucson	Citizens of Tucson
			Wildlife habitat support	Recreation land development		Arizona wildlife
Bog Hole Impoundment	Surface rangeland runoff	Class I	Wildlife habitat	Erosion control	U.S Forest Service	Citizens of Arizona
			Livestock watering	Habitat development	Coronado National Forest	Arizona wildlife
					Arizona Game and Fish Department	Santa Cruz County ranchers and landowners
Pintail Lake	Municipal sewage effluent	Class III	Wildlife habitat support	Sewage effluent disposal	U.S. Forest Service Lakeside Ranger District	City of Showlow
			Rangeland irrigation		EPA	Arizona wildlife
					City of Showlow, Arizona	Downstream population

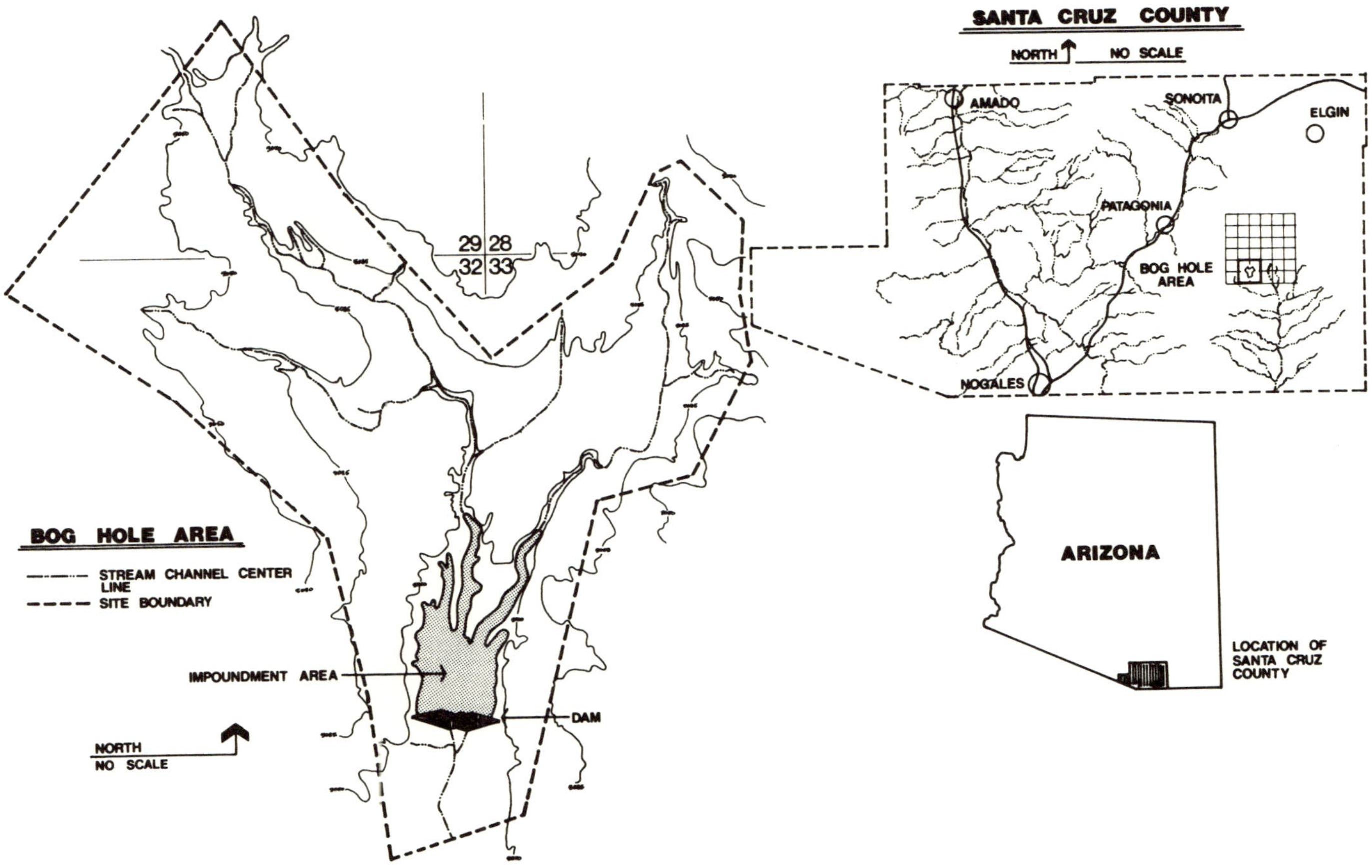

Figure 8.2. Bog Hole Water Impoundment.

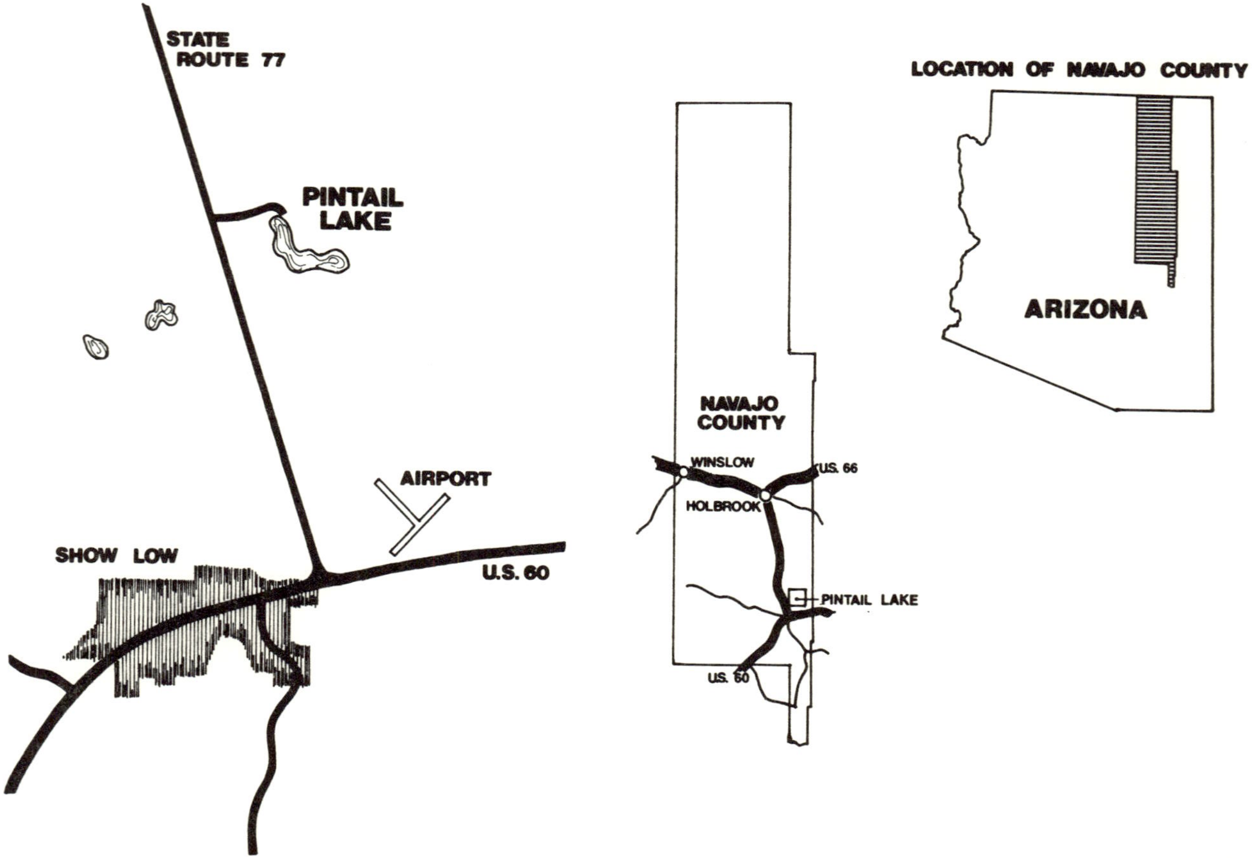
STATE
ROUTE 77
PINTAIL
LAKE
AIRPORT
SHOW LOW
U.S. 60
NAVAJO
COUNTY
WINSLOW
U.S. 66
HOLBROOK
PINTAIL LAKE
U.S. 60
LOCATION OF NAVAJO COUNTY
ARIZONA

Design Concept Summaries

There are four design concepts that contributed significantly to the overall success of these projects. These were the urban runoff–desert runoff concept (Santa Cruz River Park), the erosion control concept (Santa Cruz, Bog Hole Water Impoundment), the sewage effluent reutilization concept (Pintail Lake), and the wildlife habitat development concept (Santa Cruz, Bog Hole, Pintail Lake).

Urban Runoff–Desert Runoff. Urbanization can affect the living environment in a variety of ways. In the case of urbanization in the arid landscape, runoff is a primary example. When an arid landscape is urbanized, the immediate hydrological effect is to increase the area of low and zero infiltration capacity and to increase the speed of water transmissions in arroyos, channels, and pipelines. There is an associated decrease in drainage density or number of actual channels to carry the water and an increased sediment load.

This increased storm runoff leads to difficulties with storm drainage control, channel maintenance, stream water quality, and even groundwater recharge. Consequently, the management costs, rehabilitation costs, and the issue of safety to property and lives can mask the potential of putting to use the large volumes of water that result from these sudden runoffs.

The city of Tucson took a bold step forward when it perceived this problem as an opportunity for creating recreation and wildlife habitat sites using water control structures and a revegetation management plan. This plan perceived urban runoff and desert runoff to be a primary source of water to meet the needs of the river park. A preliminary calculation of the watershed revealed an excess of 8,700 acre-feet of runoff resulting from a 10-year storm within the urban watershed.

A comprehensive storm water detention and retention plan was created to capture and manage this resource (Fig. 8.4). The design concept identifies a series of detention ponds, infiltration beds, drainage fields, irrigation ponds, and control channels to direct and reuse the water resource along the course of the river park.

The plan called for the reconstruction of the main channel and tributary washes to detain storm runoff whether receiving a sixty- or one hundred-year flood. These in-channel detention structures are coordinated to supply adjacent retention sites with the larger-term use of excess storm water (Figs. 8.5, 8.6). Water would be applied directly through impoundment absorption or applied from remote retention ponds to recreation sites and wildlife habitat areas strategically located in-channel or adjacent to channels throughout the urban watershed. The urban river park could eventu-

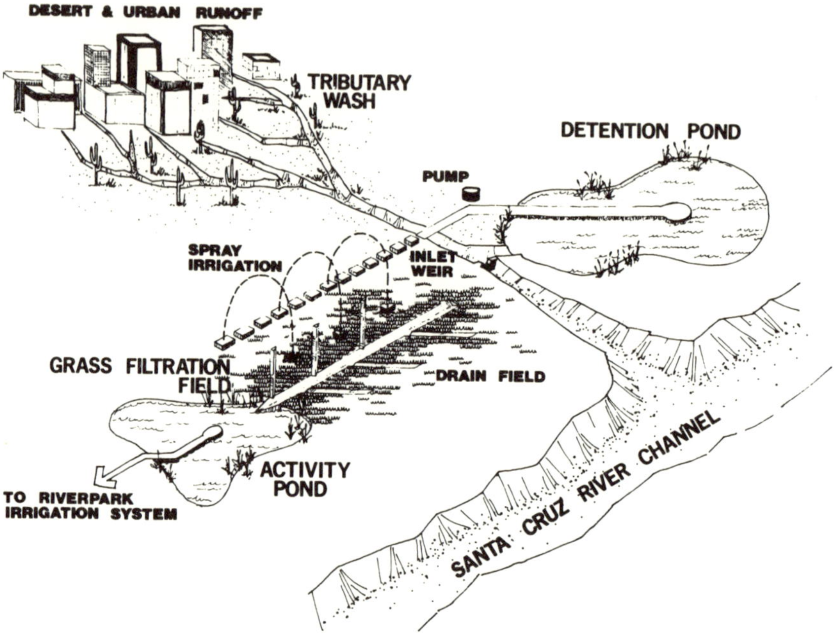

Figure 8.4. Water detention and retention plan.

ally regenerate a portion of its former desert riparian environmental character under this system. While the project is still under construction at the time of this writing, it is hoped that the concept will reestablish a vital ecosystem component to the desert environment for all to enjoy.

Erosion Control. The relative importance of runoff in any given region is affected by topography, geology, climate, soil characteristics, vegetation, and land use. In the arid climate, runoff from infrequent storms creates high flows and heavy sediment yields. Consequently erosion control was another important design concept because of the cost resulting from runoff in the arid landscape. If this erosion process can be reversed or stopped, then storm runoffs can be perceived more positively than when they create erosion problems. The Santa Cruz River Park demonstrates one attempt to reverse this process. The Bog Hole Water Impoundment represents another effort.

The extensive grama grasslands and Emory oak woodlands provide ample grazing opportunity for livestock in the Patagonia district of the Coronado

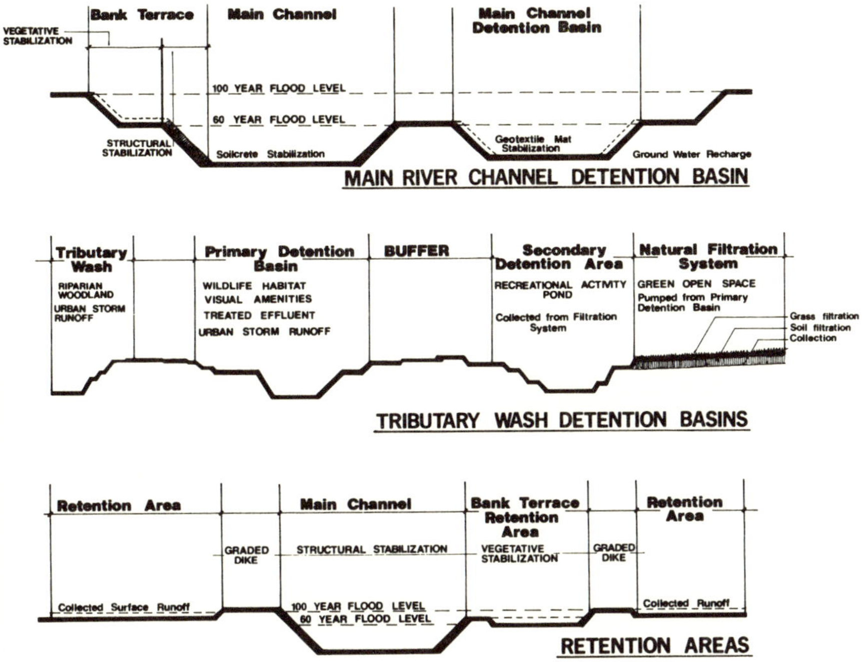

Figure 8.5. Channel design scheme for Santa Cruz River Park.

National Forest where the Bog Hole Impoundment is located. The area is incised by ephemeral meandering streams. Where stream courses occur, considerable erosional activity is taking place, presumably the result of years of grazing activity. This area has a history of grazing and mining dating back to the early Spanish period in the Southwest.

Overland flow is becoming increasingly concentrated in a decreasing number of larger drainage stream channels throughout the watershed. The Bog Hole Water Impoundment located near the headwaters of the Santa Cruz River is designed to decrease this erosional activity by decreasing volume and intensity flows along the main stem channel (Fig. 8.7). The dam impoundment has a spillway system designed to release water at reduced rates along grass-lined channels to the downstream channel. Upstream sediment traps were proposed if sediment loads caused rapid filling.

To date, the upstream channels feeding the impoundment have revegetated naturally creating their own sediment trap capability. Consequently, erosion has been greatly reduced downstream of the impoundment.

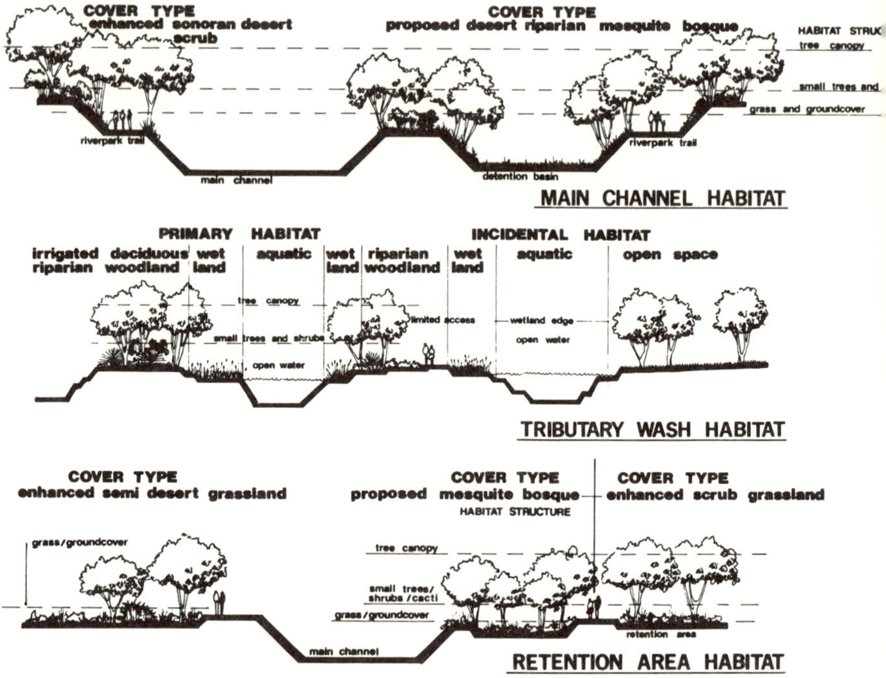

Figure 8.6. Vegetative treatment scheme for Santa Cruz River Park.

Sewage Effluent Reutilization. Sewage water contains a diverse collection of organic and inorganic materials that can ultimately affect the quality and potential uses of any existing water source that contains it. Treatment systems are designed to reduce the organic content of the wastewater through bacterial metabolism. Organic material from the wastewater provides the substrate for metabolism. Mesophilic bacteria (68–110°F/20–45°C temperature range) and cryophilic bacteria (28–86°F/ – 2–30°C temperature range) are found in most waste treatment processes. These bacteria, along with aerobic bacteria, find the Pintail Lake effluent perfectly suited to their needs. Energy is supplied by natural light and chemical reactions in the water. Such nutrients as nitrogen, phosphorous, carbon, and trace elements come from the effluent material. Consequently, a primary food chain is generated in this chemical brine.

Pintail Lake is not suitable for swimming or fishing. It is, however, ideally suitable as a waterfowl nesting and reproducing site and a rangeland

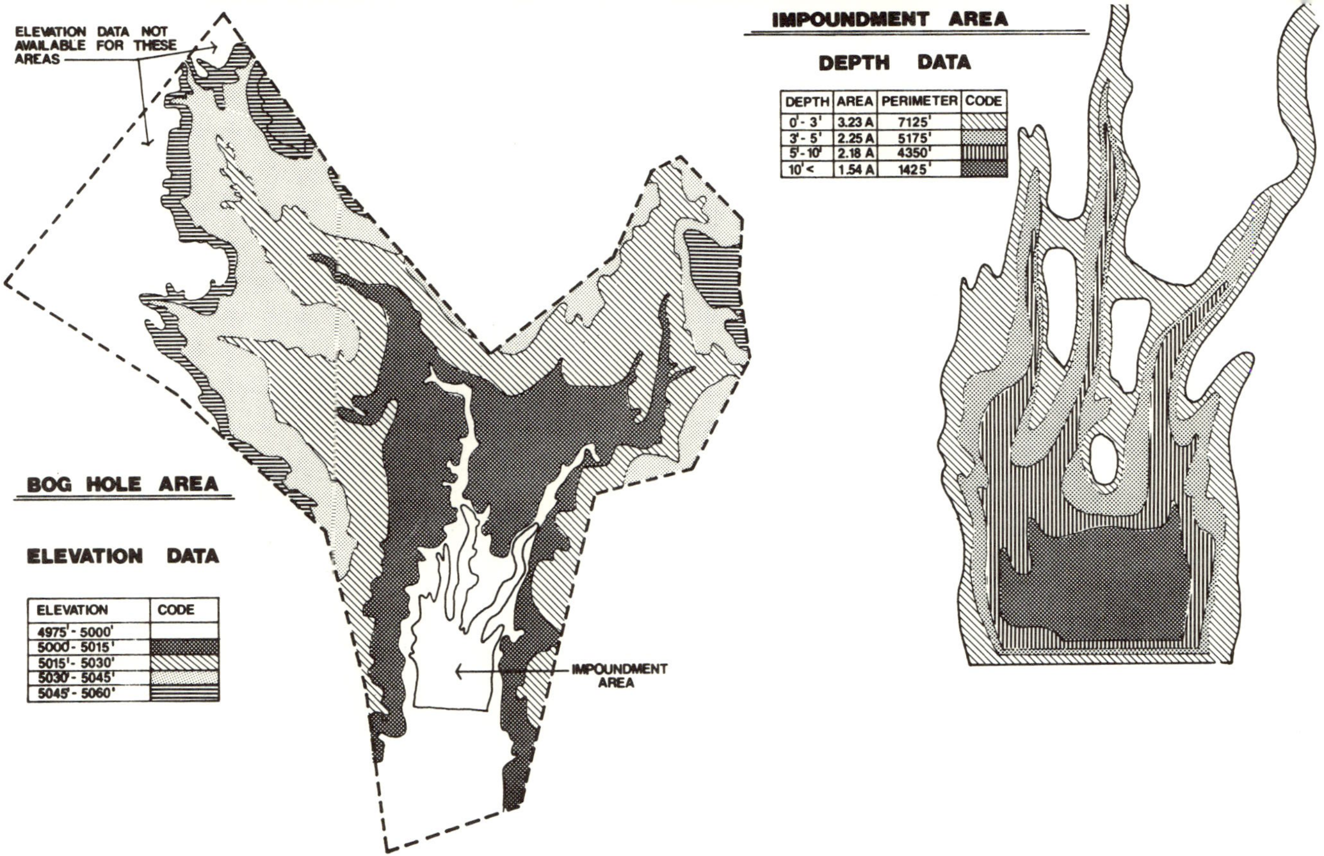

Figure 8.7. Land area analysis for the Bog Hole area.

irrigation site (Fig. 8.8). Islands provide the land cover. Aquatic plants thriving in the replenished sewage effluent provide cover and food. Pintail Lake is now a very successful non-game wildlife observatory and a seasonal waterfowl hunting area. It is also a detention impoundment that allows the surrounding ephemeral streams to remain naturally clean and clear.

Wildlife Habitat Development. Wildlife is becoming a cultural luxury. Land use patterns are developed against a background of bioclimatic, economic, and geographic dynamism. Population increases in the arid region demand careful planning. What we do with our environment will now be done by design. Wildlife will not survive unless we plan for it to be part of that design. In the future, wildlife will exist within our midst only if we plan for its presence.

All three of these projects have committed resources to develop vegetation, water, and land for the resident wildlife existing there. The Santa Cruz River Park has planned for the extensive redevelopment of riparian vegetation along its borders. Once vegetation has become reestablished, people will see songbirds, small mammals, migratory waterfowl, birds of prey, lizards, a variety of snakes, and even some larger mammals on occasion.

The Bog Hole Water Impoundment and its fenced enclosure provides aquatic, wetland, and rangeland habitats for resident wildlife, migratory wildlife, and some newly introduced species as well. Many of these species could neither exist nor thrive in this area without the year-round forage, cover, and water provided by this project.

Pintail Lake is very similar to the Bog Hole in that it provides aquatic, wetland, and rangeland species with habitat not normally available so close to human settlements. The harsh winters and arid summers normally prevent wildlife species from existing here in such concentrated numbers. Now a more permanent, high-quality habitat has been constructed through the use of an urban byproduct.

Project Evaluation Summary

Santa Cruz River Park. The Santa Cruz River Park was the most ambitious project in terms of total influenced land area (12 linear mi) and total potential managed water volume (2.8 billion gal) (Table 8.7). The project was also the most complex project to manage in terms of special interest clientele and beneficiaries. Therefore, it was also the project most likely to affect the largest segment of society with respect to this new allocation of the water resource.

Eight major planning units were identified within the linear river park.

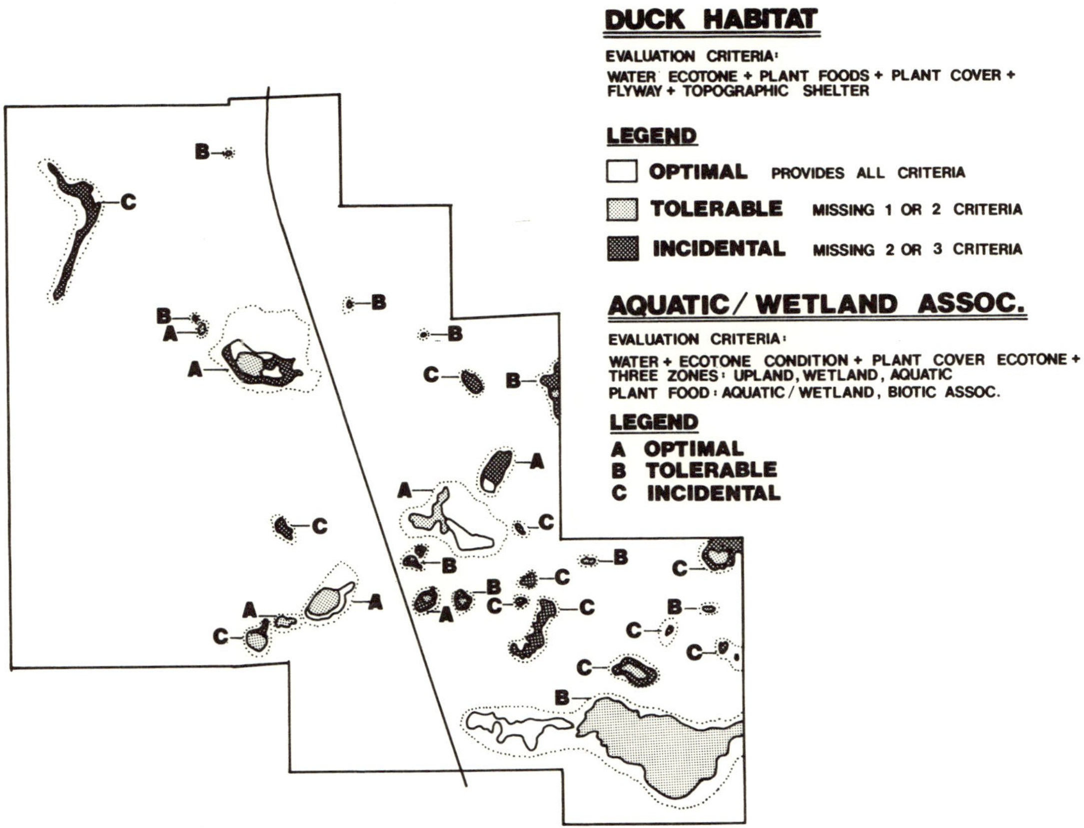

Figure 8.8. Wetland aquatic habitats, Pintail Lake.

Table 8.7. Project description summary

Project	Total Land Area	Water Volume	Water Uses	Products
Santa Cruz River Park	12 linear mi (19.3 km) of linear River Park through Tucson CBD	10-year flood event = 8,700 acre-feet or 2.8 billion gal or 749,000,000 l	Irrigation for recreation and wildlife habitat Local groundwater recharge	Desert riparian ecosystem restoration Recreation and wildlife habitat site development
Bog Hole Impoundment	198-acre (80.1 ha) exclosure 9.2 surface acres (3.7 ha) of water	49.9 acre-feet of water or 16.3 million gal or 42,960,000 l	Water for wetland and aquatic habitat Water for wildlife and livestock	Erosion control Wildlife habitat restoration
Pintail Lake	250-acre (101 ha) exclosure 57 surface acres (23 ha) of water	500,000 gal sewage effluent per day (132,000 l) 1.5 acre-feet/day inflow 1.0 acre-feet/day evaporation maximum	Sewage effluent disposal Waterfowl habitat development	Wetland and aquatic habitat restoration Waterfowl reproduction area

Each planning unit was keyed into the river park by a combination of recreation, amenity, or public service values. The project was a major element of the recreation and open space plan for the city of Tucson. Acquisition and development of the park was given a number one priority in the capital improvement program budget for special facilities through 1985. Therefore, the success of this project not only affects the immediate environment but similar open space lands throughout the city.

The specific products realized from the initial construction were the erosion control water retention element of the plan and the actual redevelopment of new lands in the drainageway for the development of recreation and wildlife habitat opportunities. The management of the storm water runoff could prove to be a major secondary benefit to the restoration potential of an all-but-lost desert riparian ecosystem in the city of Tucson. Eventually, the restoration of riparian vegetation along the river park due to storm water irrigation could prove to be the most influential element of all. Once people are convinced by successful projects that their tax dollars can make a long-term difference for both water conservation and new multiple-use land management strategies in the arid landscape, more projects are sure to follow.

Bog Hole Water Impoundment. The Bog Hole Water Impoundment was by no means any less successful in meeting its objectives. The project was modest in total land area (198 acres) and total managed water volumes (16.3 million gal) by comparison with the Santa Cruz River Park. The project, despite its remoteness, drew major attention from area landowners. This precedent-setting project demonstrated to the public that it was possible to maintain primary land user rights for leasees involved in cattle management while still committing a portion of the rangeland and the water resource to wildlife habitat restoration on the high grassland plains. Critics of this project included people who would have the public believe that multiple use of the public resource was impossible and foolhardy.

A survey of the area in which the Bog Hole was located involving some 582 km^2 identified an average water body size of 0.68 ha with a density of 0.95 water bodies/km^2 (Rodiek and McCarthy 1979). These figures represent a relatively low density of water in the region. The aquatic habitats that do exist are predominantly man-made stock tanks, canals, and ditch systems. Therefore, the wildlife and migratory waterfowl in the area are heavily dependent upon man-made systems designed for grazing and agricultural uses. It was for this reason that state and federal officials decided to create a facility that would derive its source of water from surface runoff which was not directly used by agricultural or grazing interests.

Bottom and shoreline gradients were designed to guarantee four related wetland and aquatic habitat zones. These were the marginal floodwater zone, optimal floodwater zone, open water zone, and a deep water zone. In addition to this, three islands were constructed from excavation material to guarantee protected nesting habitats.

There are at least two types of beneficiaries generated from this project. The downstream land owners and livestock managers have not experienced one major flash flood comparable to those experienced prior to the construction of the impoundment. Erosion in channels has been curtailed dramatically as well. Most important, there now exists a major reservoir for livestock and wildlife-waterfowl needs for the area.

This impoundment has evolved into a major watering hole for pronghorn antelope (*Antilocarpa americana*), mule deer (*Odocoileus hemionus*), and white tail deer (*Odocoileus virginianus*). Aquatic birds have thrived off this resource since its fill-in has been realized. Consequently, wildlife enthusiasts have been able to see wildlife activity here that has not been available for decades prior to this impoundment's completion.

Pintail Lake. Perhaps the most dramatic success can be seen on Pintail Lake. This project, including exclosure (250 acres), islands, and impoundment (57 acres) cost $145, 750. The city of Show Low and the EPA provided $91,750. The Apache-Sitgreaves National Forest contributed the land and $55,000. These costs do not reflect the actual investment since much of the success of the project hinged on the intergovernmental cooperation in terms of technical assistance, labor, and administration.

The most surprising statistics were those reflecting the reproduction of waterfowl on Pintail Lake. Piest (1982) reported 105 cinnamon teal (*Anas cyanoptera septentrionalium*), 89 mallards (*Anas platyrhynchos platyrhynchos*), 39 gadwalls (*Anas strepera*), 32 ruddy ducks (*Oxyura jamaicensis rubida*), 27 pintails (*Anas acuta acuta*), and 26 green-winged teal (*Anas crecca carolinensis*) for a total production of 318 ducklings or 5.6 per acre of surface water. Duckling production on other improved areas in the forest did not exceed 3.9 per acre and less than 1.0 per acre on most unimproved areas.

The production levels were first counted in 1979 when 7 were reported. In 1981, totals rose to 1,544 (Table 8.8). Piest reported the hatch rate was 98% for the island nests and 60% for the upland nests. Total production was an incredible 27 per surface acre of water.

Future plans for Pintail Lake include periodic drawdowns for removal of accumulated sludge and organic matter. The non-toxic material will be incorporated in the surrounding rangeland soil. The project could very likely be expanded to include several other basin areas as predictions indicate the volume of effluent to double by 1985.

Table 8.8. Number of ducklings censused on 18 wetlands in each of three years, Apache-Sitgreaves National Forest

Year	17 Wetlands	Pintail Lake	Total
1979	859	7	866
1980	677	331	1,008
1981	1,292	1,544	2,836

Source: After Piest (1982).

CONCLUSION

The Bog Hole Impoundment and the Pintail Lake projects are now in full operation. The Santa Cruz River project will take years to complete. A comprehensive assessment of the success or failure of these projects is difficult since, as mentioned earlier, we do not possess the technical means to fully appraise the actual achievements realized. Despite this reality, there are some general trends reflected in these works that merit comment. These comments will be confined to the future availability of water and its influence on growth in the arid environment, the resolution of the problem, and hopeful futures.

The past 10 years have produced only fragmented success regarding arid-land water management policy making. One reason for this is that government and research institutions supplying answers seem to be lagging behind the increased demands for water in the arid urban metropolitan centers. This situation is rapidly being reversed as evidenced by these projects. Water supplies will play an increasingly significant role in political events in the future. Two predictions can be made with some degree of confidence. First, we will face long-term increases in costs for water use. Of more significance, we will see as a result of this increased cost a different supply-demand relationship and readjusted societal values regarding water allocations. Consequently, economic growth in the arid landscape will take on new directions, new technologies, and new forms in response to this adjusted water use pattern.

The resolution of this problem has three interrelated aspects. The first is a conceptual one. In the case of water supplies in the arid landscape, our understanding as a society of the natural system and our orientation to it is fragmented and incomplete. We perceive the use of water as a given right. This right has in the past extracted no price of any significance. Consequently the use of water implied no self-restraint or collective accountability. This attitude has changed. Once an accurate popular understanding of the actual conditions is effected, the public will respond with the appropriate modified behaviors regarding water use.

Technology has insulated us from genuinely appreciating our personal, commercial, and industrial access to water. The lack of interrelatedness and coherence in the public application of our technical skills has been more often than not a great contributor to environmental degradation in the arid environment. We must now modify our competitive uses so that they fit into a more synchronized system of operation regarding water use.

Ultimately we find the physical manifestations of our environmental problem linked most clearly to our culturally defined values and beliefs. To improve our arid landscape, both people and politics must be improved. Through our governing agencies and laws, we consciously or unconsciously shape our landscapes. These three projects are evidence of a new trend.

People are adjusting their rates of water consumption. The per capita consumption of water in Tucson dropped from over 200 gal to 100 gal in the late 1970s by people simply watering half as much and at a later time of day so as to avoid high evaporation losses. The rate of depletion of the Ogallala Aquifer (west Texas) has dropped by 54% during 1979–1984 in a 15-county area. Depth-to-water level measurements made annually in a network of 948 wells in the 15-county area were used to make this observation. The basic explanation for these savings is attributed to a change in the technical manner of delivering the water supply to the field.

The future holds some significant promise with respect to more efficient use of water, better multiple-purpose benefits resulting from these uses, and the reduction of wasteful uses. The Pintail Lake project is by no means the absolute answer for wastewater disposal in the high arid grasslands. It does offer some new approaches to proper wastewater management practices and municipal cost reductions for small, growing communities. The Bog Hole Water Impoundment has not yet survived enough wet-dry cycles to prove its appropriateness. However, the rangeland condition in the exclosure and the added water supply now available have led some ranchers to conclude that certain grazing practices are exaggerating the desertification of these fragile grasslands. New rangeland research could lead this industry into a whole new way of using these grasslands on a more extended, yet more productive, rotation grazing cycle.

The Santa Cruz River Park may have the most significant impact of all three projects. Urban runoff water in the desert ecosystem has yet to be studied at any in-depth level. If the detention irrigation system shows promise, there may be a whole new potential for developing urban landscapes based on this runoff capture concept.

These projects, sponsored by local, state, and federal governmental agencies, have demonstrated that we are still able to improve upon the ways we use our water resources in the arid and semi-arid environment. These gains were made by internalizing our costs within our own human resource system. Ultimately, we must face the fact that limits must be set in the arid environment. These projects simply help us realize the progress we have made in pushing ourselves into the environment and the increased costs for remaining there.

REFERENCES

Chakroff, M. 1976. *Freshwater Fish Pond Culture Management.* Action/Peace Corps Program Training Journal Manual Series, Number 1B. N.p.

Frasier, G. W., and L. E. Meyers. 1983. *Handbook of Water Harvesting.* Agricultural Research Service; USDA Agriculture Handbook, Number 600. N.p.

Lauritzen, C. W. 1960. "Groundcovers for Collecting Precipitation." *Farm and Home Science.* 21:66–67.

Piest, L. A. 1982. *Evaluation of Waterfowl Habitat Improvements in the Apache-Sitgreaves National Forest, Arizona.* Master's thesis, University of Arizona, Tucson.

Rodiek, J. E., and M. M. McCarthy. 1979. "Riparian Wildlife Habitat Analysis in the Arid Southwestern United States." *Landscape Research.* Vol. 4, (1):4–8. Manchester, England.

Sherard, J. L., et al. 1963. *Earth and Earth-Rock Dams.* New York: John Wiley & Sons, Inc.

United Nations, ECAFE. 1962. *Proceedings of the Regional Symposium on Dams and Reservoirs.* Flood Control Series No. 21. New York.

U.S. Bureau of Reclamation. 1973. *Design of Small Dams.* 2d ed. Washington, D.C.: U.S. Government Printing Office.

U.S. Department of Agriculture. 1973a. "Hauling Water to Livestock on the Range." Technical notes, Soil Conservation Service. Salt Lake City, Utah.

U.S. Department of Agriculture. 1973b. "Wildlife Water Facility." Technical Notes, Soil Conservation Service, Ankeny, Iowa.

U.S. Department of Agriculture. 1982. *Ponds—Planning, Design, Construction, Soil Conservation Service.* Agricultural Handbook No. 590.

CONTRIBUTORS

Eric G. Bolen
Department of Range and Wildlife Management
Texas Tech University
Lubbock, Texas

William J. Clark
Wildlife and Fisheries Science Department
Texas A & M University
College Station, Texas

Gerald A. Cole (Retired)
Department of Zoology
Arizona State University
Tempe, Arizona

Paul N. Gray
Department of Range and Wildlife Management
Texas Tech University
Lubbock, Texas

David E. Jones
Arizona Cooperative Fish and Wildlife Research Unit
School of Renewable Natural Resources
University of Arizona
Tucson, Arizona

Steven J. Loring
Department of Biology
New Mexico State University
Las Cruces, New Mexico

William P. Mackay
Department of Biology
New Mexico State University
Las Cruces, New Mexico

Richard L. Noble
Department of Zoology
North Carolina State University
Raleigh, North Carolina

Kenneth G. Renard
Southwest Rangeland Watershed Research Center
2000 East Allen Road
Tucson, Arizona

Jon E. Rodiek
Department of Park Administration and Landscape Architecture
Texas Tech University
Lubbock, Texas

John L. Thames
School of Renewable Natural Resources
University of Arizona
Tucson, Arizona

Walter G. Whitford
Department of Biology
New Mexico State University
Las Cruces, New Mexico

Charles D. Ziebell
Arizona Cooperative Fish and Wildlife Research Unit
School of Renewable Natural Resources
University of Arizona
Tucson, Arizona